Computer-Assisted Structure Elucidation

Dennis H. Smith, EDITOR

Stanford University School of Medicine

A symposium sponsored
by the Division of
Chemical Information at
the 173rd Meeting of the
American Chemical Society,
New Orleans, La.,
March 23, 1977

A C S   S Y M P O S I U M   S E R I E S   **54**

AMERICAN CHEMICAL SOCIETY

WASHINGTON, D. C.   1977

Library of Congress CIP Data

Computer-assisted structure elucidation.
  (ACS symposium series; 54 ISSN 0097-6156)

  Bibliography: p.
  Includes index.

  1. Chemical structure—Data processing—Congresses.
  I. Smith, Dennis H., 1942-   . II. American Chemical Society. Division of Chemical Information. III. Series: American Chemical Society. ACS symposim series;
54.

QD471.C594   543'.08   77-24427
ISBN 0-8412-0384-9   ACSMC8   54 1–151

ACS Symposium Series

Robert F. Gould, *Editor*

FOREWORD

The ACS SYMPOSIUM SERIES was founded in 1974 to provide a medium for publishing symposia quickly in book form. The format of the SERIES parallels that of its predecessor, ADVANCES IN CHEMISTRY SERIES, except that in order to save time the papers are not typeset but are reproduced as they are submitted by the authors in camera-ready form. As a further means of saving time, the papers are not edited or reviewed except by the symposium chairman, who becomes editor of the book. Papers published in the ACS SYMPOSIUM SERIES are original contributions not published elsewhere in whole or major part and include reports of research as well as reviews since symposia may embrace both types of presentation.

CONTENTS

PREFACE

Elucidation of unknown molecular structures occupies a significant amount of the chemist's time in many areas of chemical research. A variety of physical and chemical methods are available to assist in this task. Suitably programmed digital computers are additional tools which can help to solve structural problems. If used intelligently, the speed and thoroughness of the computer can be a powerful asset to the chemist, both in decreasing time required for analysis and in ensuring that all plausible alternatives have been considered. (Computers play a critical role in x-ray crystallographic techniques also. However, such techniques were not addressed in this symposium.)

This symposium was held at a time when:

(1) Most organizations involved in chemical research have available computer systems to handle many phases of acquiring and reducing experimental data

(2) Some of the earliest methods using computers to assist in structure elucidation, e.g., library search techniques, are widely available and are incorporated into commercial systems

(3) Computer networking is making more recent problem-solving programs available to many chemists at relatively low cost

These developments have placed powerful computer systems in the laboratory for routine work and, by resource sharing via networking, have reduced the lag time for application of new computer techniques to days or weeks rather than months or years.

The participants in this symposium are using computers in several different ways to help solve unknown structures. Methodologies discussed include library search techniques, automated interpretation of data, pattern recognition, structure generation, and ranking of candidate structures based on prediction of spectroscopic or chemical behavior. It is striking to see a convergence of ideas and techniques in these seemingly diverse methods. If one views structure elucidation as a transformation of data (and representations of other information about an unknown) into representations of molecular structure which are potential solutions, then the reasons for the convergence are obvious. The computer is a powerful aide in carrying out these transformations.

I express my sincere thanks to Morton Munk for his helpful suggestions in arranging the symposium and for chairing a portion of the program. The officers of the Division of Chemical Information, especially Cynthia O'Donahue, were most helpful in assisting my efforts.

Stanford University School of Medicine   DENNIS H. SMITH
Stanford, Calif.
May 31, 1977

Computer-Assisted Structure Identification of Unknown Mass Spectra

R. VENKATARAGHAVAN, H. E. DAYRINGER, G. M. PESYNA,
B. L. ATWATER, I. K. MUN, M. M. CONE, and F. W. McLAFFERTY

Department of Chemistry, Cornell University, Ithaca, NY 14853

Mass spectrometry has become a routine technique for structure identification in a number of applications (1). Gas chromatograph/mass spectrometer/computer (GC/MS/COM) systems capable of producing a mass spectrum every second are commercially available (2). Voluminous amounts of data are generated with such systems using subnanogram amounts of sample. For full utilization of this highly specific information it is essential to employ computer techniques. Such computer-aided structure identification from mass spectrometric data has taken two distinct directions (3). The first utilizes "retrieval" systems which compare the unknown data to a library of reference spectra to report compounds with a high degree of similarity. A number of techniques have been employed for the retrieval approach (3). The second approach involves interpretive schemes that attempt to identify part or all of the unknown structure from correlations of mass spectral fragmentation behavior. Pattern recognition (4) and artificial intelligence (5) are examples of such schemes that have been employed for interpreting mass spectral data of specific classes of compounds. We will describe here a retrieval Probability Based Matching (PBM) system (6, 7) and an interpretive Self-Training Interpretive and Retrieval System (STIRS) (8 – 11) developed for the analysis of low resolution mass spectra. Both these systems are available on a computer network (TYMNET) from an IBM-370/168 computer system at Cornell University to outside users.

Probability Based Matching System

It has been shown that to increase the relevancy of information retrieved from a library of data it is essential to attach proper weighting to the contents of the system (12). The PBM system employs a probability weighting to both the mass and abundance data (6, 7). The abundance values are weighted according to a log normal distribution (13) and the masses are given a uniqueness value based on their occurrence probability

in a mass spectral data base of 18,806 different compounds (14).
The PBM system also uses a reverse search strategy, independently proposed by Abramson (15), which is valuable in identifying
components of a mixture. This technique demands that the peaks
of the reference spectrum be present in the unknown, but not that
all peaks of the unknown be present in the reference. The degree
of match of the reference to the unknown is indicated with a confidence index K, based on the statistical probability that this
degree of match occurred by coincidence; details of the method
have been described elsewhere (6, 7).

A statistical evaluation of PBM's performance was made
using "unknown" mass spectra, for each of which at least one
other spectrum of the same compound was present in the data
base. Low and high molecular weight sets, each of ~400 unknown
spectra removed at random from the data base, were run through
the PBM system, and the results evaluated using recall and
reliability as measures of performance. Recall (RC) is defined as
the number of relevant spectra actually retrieved and reliability
(RL) is the proportion of retrieved spectra which are actually
relevant. In addition to these terms it is desirable to express
the performance of automated systems in terms of false positives
(FP), the proportion of spectra predicted incorrectly (16).

$$RC = I_c/P_c \tag{1}$$

$$RL = I_c/(I_c + I_f) \tag{2}$$

$$FP = I_f/P_f \tag{3}$$

where I_c = number of correct predictions, P_c = total possible
number of correct predictions, I_f = number of false predictions,
and P_f = total possible number of false predictions. At the 50%
recall level the reliabilities for the low and high molecular weight
sets were 65% and 42%, counting as correct only predicted structures which are identical to the unknown. Invariably retrieval
systems predict similar structures in addition to the identical
structure. In the evaluation of PBM results four classes of
similarity were defined: I, identical compound or stereoisomer;
II, class I or a ring position isomer; III, class II or a homolog;
IV, class III or an isomer of class III compound formed by moving
only one carbon atom. It was found that when class IV type compounds were accepted as correct predictions the reliability of the
system increased to 95% at the same recall level.

Recently, it has been found that the performance of PBM
for the identification of components in a mixture can be enhanced
(17) by incorporating a spectrum subtraction procedure similar to
the one proposed by Hites and Biemann (18). The method subtracts
the reference compound matched by PBM with the highest confidence index (or any other in the list of predicted spectra) from the
unknown spectrum and matches the residual peaks against the

reference file by PBM. This operation is particularly valuable for identifying a minor component missed by the reverse search procedure when there is substantial overlap in the spectra of the major and minor component, or when amount of the latter falls outside the limits set for "percent component" or "percent contamination".

Self-Training Interpretive and Retrieval System

The STIRS system is an interpretive scheme that trains itself for the identification of different structural features in an unknown by utilizing specific classes of mass spectral data (8). Table I shows the fifteen data classes used; although these have been selected for their structural significance, there are no predesignated correlations of specific spectral data with corresponding structures. For each unknown spectrum the system matches its data in each class against the corresponding class data of all reference spectra and computes a match factor (MF) indicating the degree of similarity. In each data class the fifteen reference compounds of highest MF values are saved. If a particular substructure(s) is found in a significant proportion of these compounds, its presence in the unknown is probable. Absence of a substructure is not predicted, as the mass spectral features of one substructure can be made negligible by the presence of a more powerful fragmentation-directing group. The data base for the system includes information from 29,468 different organic compounds containing the common elements H, C, N, O, F, Si, P, S, Cl, Br, and/or I. All structures of these compounds have been coded in Wiswesser Line Notation (WLN) to facilitate computer handling of structure data.

To utilize the information provided by the STIRS system, the results for each data class are examined and the common structural features identified. To aid this process, in a recently implemented system (9), the computer examines the data for the presence of 179 frequently found substructures (19). The probability for the presence in the unknown of each substructure is predicted using a random drawing model. Knowing the frequency of occurrence of a specific substructure in the file, this method indicates the probability that the prediction of its presence in the unknown occurred at random. From this probability the confidence for each prediction is calculated. For example, in the STIRS data base the phenyl substructure is found to be present in 28% of the compounds. Statistically on the average this substructure would occur in 4 of any 15 compounds in the data base, including the top 15 compounds selected in a STIRS data class. On the other hand if phenyl is found in 10 of the 15 compounds, the probability that this occurred by chance is only 1 in 113, so that the confidence in the phenyl prediction is >99%, or a false positives value of <1%.

Table I. Mass Spectral Data Classes Used in STIRS

Data Class	Description, maximum number of peaks	Range of mass or mass loss
1	Ion Series (14 amu separation)	<100
2-4	Characteristic ions	
2A	Four even-mass, four odd-mass	6-88
2B	Eight	47-102
3A	Seven	61-116
3B	Seven	89-158
4A	Six	117-200
4B	Six	$159-(M-1)^+$
5-6	Primary neutral losses	
5A	Five	0-2, 15-20, 26-53
5B	Five	34-75
6A	Five	59-109 (MW $\geq$175)
6B	Five	76-149 (MW $\geq$250)
5C	Five	16-20, 30-38, 44-51, 59-65, 72-76
6C	Five	26-28, 39-42, 52-56, 62-70, 80-84
7, 8	Secondary neutral losses from most abundant odd-mass (MF7) and even-mass (MF8) loss	<65
11	Overall match factors	
11.0	MF11.1 + MF11.2	
11.1	2A + 2B + 3A + 3B + 4A + 4B	
11.2	5A + 5B + 6A	

The system has been extensively tested for each of the 179 substructures by selecting 373 compounds at random from the data base (every 50th compound in the Registry data) (20). If the data set did not contain at least 30 compounds with a particular substructure, the required additional compounds were selected at random that contained the substructure. If fewer than 30 compounds with a given substructure were available, all of them were selected. System performance in each data class was evaluated by computing recall and reliability terms for each substructure. In contrast to equation 2, the reliability term in this case included a false positive factor, being set equal to $RC/(RC + FP)$, such that the values reflect the system performance averaged for compounds containing and not containing the substructure. This reliability term led to substantial confusion, so that we now feel that it is better to report performance of the system in terms of recall and false positives (16), as discussed for PBM (equations 1 and 3).

Analysis of the data shows that although individual data classes are good for specific substructure identification, the best performance is found in the overall match factor (Table I) results. This is due to the fact that the overall match factor data combines the information derived from the individual data classes. The overall match factor, MF11.0, which combines ion series, characteristic ions, and neutral loss data has been found to give the most reliable information on the different substructure possibilities in an unknown compound. For the 179 substructures tested, the MF11.0 gave a recall of 49% at 1.9% false positive level. A number of improvements have been made to the characteristic ion data classes (10) and the primary neutral losses (11); the overall match factors MF11.1 and MF11.2 have been found to give an average recall of 47% and 32.1%, respectively, at $\leq 2\%$ false positives level.

For the different substructures tested the recall values varied widely. This is due to the variation in the prominence of mass spectral characteristics of different substructures that were tested. For example, complex substructures such as adamantyl (RC = 67%), benzphenanthrene (RC = 89%) and oxazole (RC = 83%) have been found to give high recall values whereas considerably simpler substructures such as carbonyl (RC = 31%) did not. This is not a limitation on the capabilities of the system but rather a characteristic of the mass spectral behavior of the different substructures. In the case of a carbonyl group there is no distinct peak associated with the substructure, but the visibility of the group increases in carbonyl-containing substructures such as acetyl and benzoyl. Although the overall "characteristic ions" match factor, MF11.1, has been found to give better recall values for most substructures than does MF11.2, the overall neutral loss match factor, for certain individual substructures the recall values of MF11.2 are far superior (11), complementing the ion series and characteristic ions data. This effect is apparent in the

MF11.0 results which combine the capabilities of all these data classes.

The 179 substructures automatically sought by STIRS were selected because they appear in the dictionary of frequently found substructures (19), not because of their mass spectral behavior. We are currently developing a computer-assisted procedure to make extensive mass spectral correlations (21) of the STIRS data base. This study should provide a set of substructures that produce the most characteristic mass spectral data, thus taking full advantage of the information available in the data base. This procedure initially selects significant peaks from each spectrum using the criteria developed for the PBM system. The peaks are then assigned all possible elemental compositions that are consistent with the molecular formula of the parent compound. Each composition is then assigned possible structures that are compatible with parent structures by using a graph walk algorithm. Each structure is then assigned a score based on mass spectral fragmentation rules. At this stage structures and compositions will be examined manually for consistency and correctness of assignment. The validated data will then be tabulated in terms of mass, elemental compositions, and substructures also indicating their significance within the data base.

Even with testing the 15 STIRS-selected compounds for the 179 substructures, it would be helpful if the computer could compare these structures to identify their common maximal substructures. We have written a computer program to accomplish this task (22). This involves the comparison of two structures at a time, coding them in a Wiswesser connection table; compatibilities between nodes of each structure are established taking into account the connectivity relationship of the nodes. The Compatibility Table generated by this algorithm is then examined to locate all the maximal common substructures. This process is repeated for all pairs of structures in the 15 compounds of each data class, and the most frequently found substructures are reported. The substructure information derived from the above procedures could then be used in the constrained structure generator (CONGEN) program, described by Carhart, et al., at this symposium, to build all possible structures for the unknown using the molecular weight information. We have also designed an algorithm to determine the molecular weight from the mass spectral data even when the molecular ion is absent in the spectrum.

The retrieval and interpretive systems described here are designed to be used as an aid to the interpretation of mass spectral data. If the spectrum of the unknown is matched sufficiently well by a library spectrum, then the structure identification is greatly simplified. If a spectrum of the unknown is not in the library, STIRS can suggest possible substructures to speed the elucidation of the total structure. Both the PBM and STIRS systems are small enough to run on a laboratory computer system, requiring under one minute to be run in our DEC PDP-11/45 system. The programs are also available for outside users through TYMNET from an IBM-370/168 computer at Cornell University (23). The increased usage of these computer-assisted structure identification procedures over the computer network in the last two years has already demonstrated the value of such systems. The anticipated growth in the application of GC/MS to important problems suggests that further development in this area is essential.

Applications

PBM Spectrum Subtraction. Brenner and Suffet (24) have recently reported comparative studies of various computer systems for mass spectral identification as applied to trace organic analysis of natural waters. The spectrum of an unknown mixture which they determined to contain bis(2-chloroethyl)ether and C_3-benzene isomers is given in Table II.

Brenner and Suffet report (24) that the Biemann/MIT matching system (18) indicated the presence of C_3-benzenes, although with a relatively low probability (Table II). Similarly, STIRS only indicated the aromatic constituents, with 11 of the 15 compounds selected by the overall match factor containing a benzene ring. Only one of the top 15 compounds was dichlorinated; the performance of STIRS is generally much better for the mass spectra of relatively pure compounds. In contrast to the results from the MIT and STIRS system, the PBM examination of the unknown spectrum found the other component, bis(2-chloroethyl)ether (Table III); apparently no C_3-benzene spectra were retrieved because of the restrictions on % contamination and % component (14). However, at the conclusion of the PBM run the best-matching reference spectrum is automatically subtracted from the unknown spectrum, and when this was run, all of the best-matching spectra found by PBM were of C_3-benzenes (Table III).

The best-matching reference spectrum of this run is again subtracted, and the resulting residual of the residual does show significant peaks indicative of one or more additional components, such as at m/e 57, 71, and 85 which do not arise from either the bis(2-chloroethyl)ether or C_3-benzenes, plus peaks which could arise from these components which have not been completely subtracted. The latter information apparently is overly confusing for PBM, as the compounds retrieved on running the residual of the residual are not particularly logical. The best-matching compound, with K = 54** and ΔK = 28, is 2-keto-3-methylvaleric acid methyl ester, and all retrieved spectra showed % contamination $\geq$50%. We find for a variety of other unknown spectra that this is generally the case; thus PBM examination should be restricted to just the first residual spectrum.

STIRS Results on a River Water Extract. The spectrum shown in Figure 1 was submitted to STIRS as an unknown at a time

Table II. Unknown from the Philadelphia Drinking Water Supply

m/e	Rel. Abund.	m/e	Rel. Abund.	m/e	Rel. Abund.	m/e	Rel. Abund.
27	53	51	10	71	7	99	1
29	33	52	3	74	1	100	1
30	1	53	4	75	1	102	1
31	5	55	4	77	18	103	7
34	2	56	5	78	5	104	2
38	2	57	63	79	10	105	100
39	17	58	10	84	2	106	13
40	8	59	1	85	33	111	1
42	9	61	1	86	2	115	2
41	30	62	3	89	1	117	2
43	30	63	53	91	7	119	5
44	2	64	2	92	1	120	34
45	3	65	18	93	60	121	3
49	3	66	1	94	2	127	1
50	5	69	4	95	17	142	3
		70	4	98	2		

Matches from Biemann/MIT Search on MSSS:

Name	Similarity Index
3-(2'-methylphenyl)-pentane	0.228
4,6,8-Trimethyl-2-methoxy-1-azocine	0.212
1,3,5-Trimethylbenzene	0.185
1,2,4-Trimethylbenzene	0.183
1,2,3-Trimethylbenzene	0.176
Nezukone	0.164
sec-Butylbenzene	0.162
1-Benzoyl-4-methylpentane	0.157
1-Methyl-2-n-propylbenzene	0.153
2-Phenylhexane	0.151

Table III. PBM Results on Unknown and Residual Spectra from
  Table II

Compound	K	ΔK	% Contamination	% Component
bis(2-Chloroethyl)ether	62**	37	24%	53%
2-(2-Chloroethoxy)ethanol	53***	46	26%	64%
bis(2-Chloroethyl)ether	51+	45	29%	53%
bis(2-Chloroethyl)ether	47+	50	30%	38%

First reference spectrum subtracted, residual spectrum run on
  PBM:

Isopropylbenzene	73+	11	34%	77%
Isopropylbenzene	72+	10	34%	91%
Isopropylbenzene	72+	9	34%	83%
Isopropylbenzene	71+	10	43%	72%
1-Methyl-2-ethylbenzene	61**+	20	36%	74%

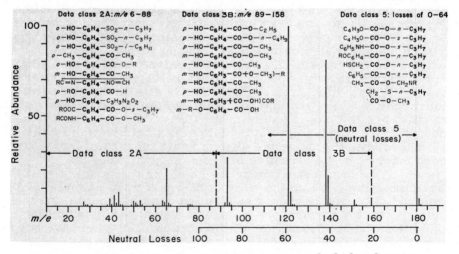

Figure 1. STIRS results for the mass spectrum of n-propyl p-hydroxybenzoate

when a reference spectrum of this compound, n-propyl p-hydroxybenzoate, was not in the data base. Results for three of the 15
data classes illustrate the "self-training" feature by which STIRS
indicates structural features of the unknown. Data class 2A
utilizes the largest peaks in the low mass region of the spectrum
(m/e 6-88); these fragment ions are more often formed by secondary reactions of higher energy requirements, and so are indicative
of gross, rather than specific, structural features. Thus all of
the spectra found of highest MF2A values contained a phenyl
group, although the phenyl rings in these compounds contain a
rather wide variety of substituents. The experienced mass spec-
trometrist probably would have inferred the presence of phenyl
from the "aromatic ion series" in this region; however, STIRS was
not trained specifically to recognize these features, but instead
indicated the presence of phenyl by finding that such compounds
matched these data the most closely.

Data class 3B covers a higher mass range, whose fragment peaks should be indicative of more specific structural
features. Again all compounds of highest MF3B values contain
the phenyl group, but almost all of them also contain an aryl
hydroxy group (not ortho) and a carbonyl. Note that the latter is
contained in carboxyl, ester, and keto functionalities; because
STIRS is designed to provide positive information, data class 3B
thus indicates the presence of HO-phenyl-CO-.

Data class 5 employs "neutral loss" information, the
differences in mass between the observed fragment ion and the
molecular ion, which in this case is assumed to be m/e 180.
Cleavage of the molecular ion gives two fragments, only one of
which holds the positive charge, and thus the neutral lost generally contains the more electronegative functionalities. Illustrat-
ing this, when the masses representing the most common neutral
losses of this unknown were matched against the whole reference
file, the highest MF5 values were found to be mainly propyl
esters. To reiterate, STIRS was not preprogrammed to recognize
propyl esters from their common losses of 41, 42, and 59 mass
units; STIRS in effect trains itself to recognize the propyl ester
functionality by finding that these data of the unknown were
matched best by propyl esters in the file. Note also that the compounds found by MF5 did not contain a particularly significant
number of phenyl groups; the different data classes of STIRS have
been selected to be sensitive to different functionalities.

STIRS has been designed as an aid to the interpreter; if the
interpreter now adds up the mass of di-substituted phenyl (76),
hydroxyl (17), and propyl ester (87), he can note that the sum
corresponds to the supposed molecular weight, 180, indicating
that all of the functionalities of the unknown molecule have been
identified by these three data classes of STIRS.

STIRS: Unknown Terpene. The mass spectrum of
12β-acetoxysandaracopimar-15-en-8β-,11α-diol was examined by

STIRS, omitting all spectra of this compound from the reference file. The nine structures of highest "overall match factor" (MF11.0) values are shown in Figure 2.

If the identity of this molecule had been totally unknown to the interpreter, these MF11.0 selections should have indicated at least the general structural features of the molecule to the interpreter. Thus all of the compounds of Figure 2 have either three or four fused rings and all have the three fused six-membered rings that are actually present in the unknown. The four tricyclic compounds closely resemble the correct structure in having methyl groups in the 4, 4, 10, and 13 positions, hydroxy at 8, and vinyl at 13. Note that three of the steroids contain a 5-hydroxy group, which can be viewed as corresponding to the correct 8-hydroxy position by "flipping" the structures, with their acetoxy groups then at least present in the ring corresponding to the ring containing the acetoxy group in the unknown. The presence of hydroxyl and acetoxy groups are indicated by the fact that eight of the nine compounds contain hydroxyls and seven contain acetoxy groups; only two contain more than one hydroxyl group, while none contain more than one acetoxy. However, the compound does give a molecular ion, so that it should be possible for the interpreter to infer correctly that the unknown contains one acetoxy and two hydroxyl groups after deducing the tricyclic system with the other substituents. Also, the steroid selected as the seventh compound has a 4-gem-dimethyl group. For this unknown thus STIRS can give fair confidence in all of the structure assignments except the position of the acetoxy and one of the hydroxy groups; there is even some indication of their positions, as in the majority of selected structures of Figure 2 these substituents are on the exterior ring bearing the bridgehead hydroxyl.

PBM/STIRS Examination of Unknown Spectra of Fatty Acid Esters. In an early classic case of natural product structure determination by mass spectrometry (25) a compound isolated as the methyl ester from butterfat was identified to be methyl 3,7,-11,15-tetramethylhexadecanoic acid. The original published (25) spectrum (omitted from the reference file) was run through PBM and STIRS to give the results shown in Table IV.

PBM correctly identified the compound as methyl phytanoate, retrieving the two reference spectra of this compound in the PBM reference file; note that the third selection is a much poorer match. The substructures identified by STIRS MF11.0 and 11.1 are correct, although the acetate substructure indicated by MF11.2 is not (Table IV). The best-matching compounds found by STIRS MF11.0 are all methyl esters of long-chain fatty acids, and all but one has a methyl group in the three position. The positions of the other methyl groups were not found by STIRS, consistent with the rather small effect of such methyl groups on the mass spectra.

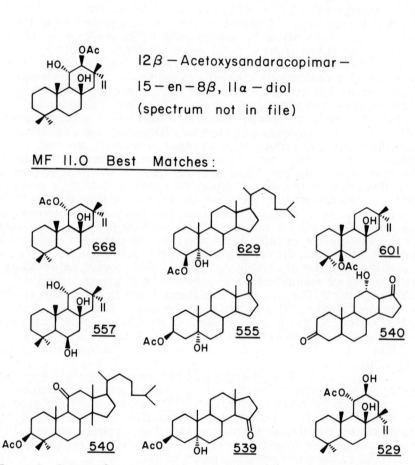

12β — Acetoxysandaracopimar —
15 — en — 8β, 11α — diol
(spectrum not in file)

MF 11.0 Best Matches:

Figure 2. Best-matching compounds and their MF11.0 values found in the STIRS examination of 12β-acetoxysandaracopimar-15-en-8β,11α-diol

Table IV. Unknown Butterfat Methyl Ester (not in file)

<u>PBM</u>:   <u>K</u>   <u>ΔK</u>

Phytanic Acid Methyl Ester   101+   0
Methyl Phytanoate   98+   2
Isopropylidene batyl alcohol   57**   43

<u>STIRS</u>: Overall Match MF11.0

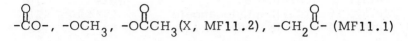

-CO-, -OCH$_3$, -OCCH$_3$ (X, MF11.2), -CH$_2$C- (MF11.1)

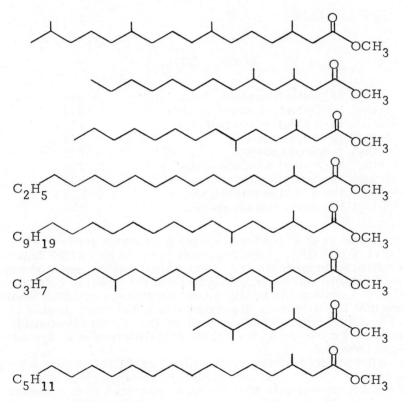

As a further test of the sensitivity of STIRS and PBM for such structural differences, the spectrum of methyl 10-methyl-nonadecanoate was run as an unknown without removing its spectrum from the reference file. Table V shows, as expected, that

Table V.   Unknown, Methyl 10-methylnonadecanoate (spectrum in file)

PBM:	$\underline{K}$	$\underline{\Delta K}$
Methyl 10-methylnonadecanoate	159+	0
Methyl n-eicosanoate	102+	10
Methyl 16-methylheptadecanoate	84**	28
Methyl 14-methylhexadecanoate	83***	29

STIRS, MF11.0:

$$\overset{\overset{\textstyle O}{\|}}{-C}O-, \ -OCH_3$$

Methyl 10-methylnonadecanoate	919
Methyl 10-D-methyloctadecanoate	844
Methyl 11-methylnonadecanoate	774
Methyl n-docosanoate	749
Methyl n-tricosanoate	745
Methyl 17-methyloctadecanoate	734
Methyl 16-methylheptadecanoate	733
Methyl 13-methylpentadecanoate	733
Methyl 15-methylhexadecanoate	726

this spectrum is then retrieved as the best match for both PBM and MF11.0 of STIRS. Note that both the PBM and STIRS runs have retrieved long-chain fatty acid methyl esters, most of which contain a single methyl group substituted far from the ester functionality. Locating the position with confidence by STIRS would require that the reference file contains a substantial number of such 10-methyl derivatives, as well as that the exact position of the methyl group produces substantial differences in the mass spectral behavior.

The MS fragmentation-directing capability of the carbonyl group is well known, and its position in an alkyl chain should have a much greater effect on the mass spectrum than a methyl group. Table VI shows the results of the PBM/STIRS examination of the mass spectrum of methyl 3-oxononadecanoate as an unknown. Now the five best matches retrieved by STIRS MF11.0 are all methyl esters of long-chain fatty acids with a carbonyl group in the three position.

Table VI. Unknown, Methyl 3-oxononadecanoate (spectrum in file)

PBM: K   ΔK

Methyl 3-oxononadecanoate   138+   0

STIRS, MF11.0:

$$-\overset{O}{\overset{\|}{C}}O-, \quad -\overset{O}{\overset{\|}{C}}CH_2\overset{O}{\overset{\|}{C}}-, \quad -CH_2\overset{O}{\overset{\|}{C}}CH_2\overset{O}{\overset{\|}{C}}-, \quad -CH_2\overset{O}{\overset{\|}{C}}CH_2-$$

Methyl 3-oxononadecanoate	878
Methyl 3-oxoheptadecanoate	837
Methyl 3-oxooctadecanoate	817
Methyl 3-oxotetracosanoate	785
Methyl 3-oxopentadecanoate	655

As a final example, the results of examining the mass spectrum of glyceryl tridodecanoate as an unknown are shown in Table VII. Although the unknown spectrum was not in the file,

Table VII. Unknown Glyceryl Tridodecanoate (not in file)

PBM: K   ΔK

Glyceryl tridodecanoate	120*	0
Vinyl dodecanoate	29***+	88

STIRS, MF11.0:

Glyceryl tridodecanoate	725
Glyceryl tritetradecanoate	596
Glyceryl-2-laurate-1,3-distearate	582
N-trifluoroacetyl-N-dodecyldodecanamide	582
Glyceryl trioctadecanoate	573
L-1,2-dilauroylglycerylphosphoroylcholine	567
Glyceryl-1-palmitate-2,3-distearate	560
Glyceryl-2-laurate-1,3-dipalmitate	552
1-Stearo-2,3-dimyristin	551
2-Stearo-1,3-diacetin	547
Glyceryl trioctadecanoin	543
Glycerol-1,3-di octadecanoate	542
Glyceryl-2-myristate-1,3-distearate	535

PBM retrieved another spectrum of this compound as an excellent match, with K and ΔK values far superior to that of the only other retrieved spectrum, that of vinyl dodecanoate. The best-matching compounds retrieved by STIRS MF11.0 were, with only one exception, all glyceryl esters containing fatty acid residues. The dodecanoate (lauric) acid moiety is found in five of the eight best-matching compounds, which should at least provide some help to the interpreter.

Practical Use of PBM and STIRS. Our main incentive of making these systems available on the Cornell computer over the TYMNET system (23) was to gain feedback on their application to real unknowns. The authors would especially appreciate receiving spectra on which PBM or STIRS apparently did not give satisfactory results. It has been our experience that the best way to understand the utility, and the shortcomings, of PBM as a retrieval system, and of STIRS as an interpretive system, is to run spectra of compounds of direct interest to the particular user. Because STIRS has been designed as an aid to the interpreter, the interpreter's knowledge of (and interest in) the sample and its chemistry should maximize the information which can be gained from examination of the STIRS results. Using these results efficiently does require experience. Suggestions as to better use of these results, or improvements to either the PBM or STIRS system, will be welcomed by the authors.

Literature Cited

1. Waller, G. R., "Biochemical Applications of Mass Spectrometry", Wiley-Interscience, New York, 1972.
2. McFadden, W. H., "Techniques of Combined Gas Chromatography/Mass Spectrometry", Wiley-Interscience, New York, 1973.
3. Pesyna, G. M. and McLafferty, F. W., in "Determination of Organic Structures by Physical Methods", pp. 91-155, Nachod, F. C., Zuckerman, J. J., and Randall, E. W., Eds., Vol. 6, Academic Press, New York, 1976.
4. Isenhour, T. L., Kowalski, B. R., and Jurs, P. C., Critical Review Anal. Chem., (1974), 4, 1.
5. Smith, D. H., Buchanan, B. G., Engelmore, R. S., Adlercreutz, H., and Djerassi, C., J. Am. Chem. Soc., (1973), 95, 6087.
6. McLafferty, F. W., Hertel, R. H., and Villwock, R. D., Org. Mass Spectrom., (1974), 9, 690.
7. Pesyna, G. M., Venkataraghavan, R., Dayringer, H. E., and McLafferty, F. W., Anal. Chem., (1976), 48, 1362.
8. Kwok, K.-S., Venkataraghavan, R., and McLafferty, F. W., J. Am. Chem. Soc., (1973), 95, 4185.

9. Dayringer, H. E., Pesyna, G. M., Venkataraghavan, R., and McLafferty, F. W., Org. Mass Spectrom., (1976), 11, 529.
10. Dayringer, H. E. and McLafferty, F. W., Org. Mass Spectrom., (1976), 11, 543.
11. Dayringer, H. E. and McLafferty, F. W., Org. Mass Spectrom., (1976), 11, 895.
12. Salton, G., "Automatic Information Organization and Retrieval", McGraw-Hill, New York, 1968.
13. Grotch, S. L., 17th Annual Conference on Mass Spectrometry, Dallas, May, 1969, p 459.
14. Pesyna, G. M., McLafferty, F. W., Venkataraghavan, R., and Dayringer, H. E., Anal. Chem., (1975), 47, 1161.
15. Abramson, F. P., Anal. Chem., (1975), 47, 45.
16. McLafferty, F. W., Anal. Chem., accepted.
17. Atwater, B. L., McLafferty, F. W., and Venkataraghavan, R., in preparation.
18. Hites, R. A. and Biemann, K., Adv. Mass Spectrom., (1968), 4, 37.
19. "ISI Chemical Substructure Dictionary", Institute for Scientific Information, Philadelphia, Pennsylvania, September, 1974.
20. Stenhagen, E., Abrahamsson, S., and McLafferty, F. W., "Registry of Mass Spectral Data", Wiley-Interscience, New York, 1974.
21. McLafferty, F. W., "Mass Spectral Correlations", Advances in Chemistry Series No. 40, American Chemical Society, Washington, 1963.
22. Cone, M. M., Venkataraghavan, R., and McLafferty, F. W., J. Am. Chem. Soc., submitted.
23. Office of Computer Services, Cornell University, Ithaca, New York 14853.
24. Brenner, L. and Suffet, I. H., Environ. Sci. Health, Part A, in press.
25. Ryhage, R. and Stenhagen, E., in "Mass Spectrometry of Organic Ions", p. 417, McLafferty, F. W., Ed., Academic Press, New York, 1963.

Acknowledgment

We are grateful to Professor I. H. Suffet of Drexel University for helpful discussions concerning the application of PBM and STIRS to actual problems, to the National Institutes of Health (grant 16609) and to the Environmental Protection Agency (grant R804509) for financial support of this work, and to the National Science Foundation for partial funding of the DEC PDP-11/45 computer used in this research.

2

Identification of the Components of Complex Mixtures by GC–MS

J. E. BILLER, W. C. HERLIHY, and K. BIEMANN

Department of Chemistry, Massachusetts Institute of Technology, Cambridge, MA 02139

It has been well-known for many years that Gas Chromato-graphy-Mass Spectrometry is a powerful tool for the qualitative identification of the components of complex mixtures. It has also been well-established that the computer is needed in all aspects of this process beginning with the initial acquisition of the data from the laboratory instrument, the processing and "crunching" of the data, and presentation of the data to the human interpreter. In recent years, techniques to improve and enhance the data, as well as the many methods of interpretation and identification have also fallen almost totally into the province of the computer.

Our laboratory has been intimately involved in the develop-ment and use of many of these techniques. Our needs are somewhat unique in the very large number of chemical problems from very diverse sources which require identifications or structure deter-minations. We currently acquire and analyze approximately one-half million spectra a year from such varied sources as drugs in body fluids, geochemical extracts, metabolism studies, organic synthetic studies, and many others. We have even had to extend the capability of our data enhancement and interpretive algo-rithms and routines to the analysis of spectra returned from the GC-MS instruments in the two Viking landers on the surface of Mars (1).

The need to analyze and interpret such a large volume of data has many implications. The most important requirement is a fully automated system capable of maximum performance and efficiency from end-to-end. This means that the GC-MS instrumen-tation must produce the best quality data possible under the fast scan conditions necessary for reasonable resolution of complex mixtures. The man-machine interface must be simple and human-engineered to limit the probability of operator error; the computer-instrument interfaces (both data and control) have to be accurate and highly reliable. The basic processing of the data must be efficient, fast, economical and complete, and the presen-tation of such vast amounts of information must be accomplished

18

in an equally efficient, yet convenient form. Perhaps most importantly (and certainly of the greatest complexity), the computer must be able to purge the spectra of background and other unavoidable interferences so that it can reliably interpret the spectra to accomplish the primary task of the identification of the components of interest.

In addition, the routine analyses of such large amounts of data require practical approaches to the design of routines to both enhance and interpret spectra. Very elegant and time-consuming algorithms which improve the results by some small fractional margin towards perfection at the expense of either large computer power or time are obviously not useful or appropriate. In a similar way, the approach to the presentation of the data in such huge quantities necessarily precludes impressive but impractical interactive graphic systems which require inordinate amounts of both human and computer time.

We feel that we have developed a practical system able to meet all of these needs. We would like to briefly describe how this has been accomplished, and in addition, describe a new and powerful technique for the automated and intelligent identification of oligopeptides in complex mixtures in support of our work on protein sequencing.

A Comprehensive System for the Analysis of Complex Mixtures

As mentioned above, the first necessary component in a balanced system for the identification of the constituents of a complex mixture is the source of the data. The GC-MS instrument must be designed and optimized for fast, continuous acquisition of data over the entire GC analysis. The scan function must be accurate and reproducible, the data must be acquired with high sensitivity, low electronic noise, and wide dynamic range to accomodate the large concentration differences common in complex organic mixtures.

This is accomplished in our laboratory on a highly modified Hitachi RMU-6L Mass Spectrometer interfaced to a Perkin Elmer 990 Gas Chromatograph via a Watson-Biemann fritted glass separator (2). Almost all of the RMU-6L electronics have been replaced with reliable solid-state circuitry, and the magnet is controlled digitally to produce a fast, reproducible scan. The output of a solid-state electrometer is digitized at three separate gain levels to achieve a very accurate signal with large dynamic range, high signal-to-noise ratio, and high sensitivity (3). The system is diagrammed in Figure 1.

The data processing includes the normal mass-intensity reduction, and in addition, the routine generation of all mass chromatograms (3,4). To print or plot this amount of information in the conventional ways for even a single sample would be very time-consuming and difficult. To solve the data presentation problem, all spectra and mass chromatograms are directly micro-

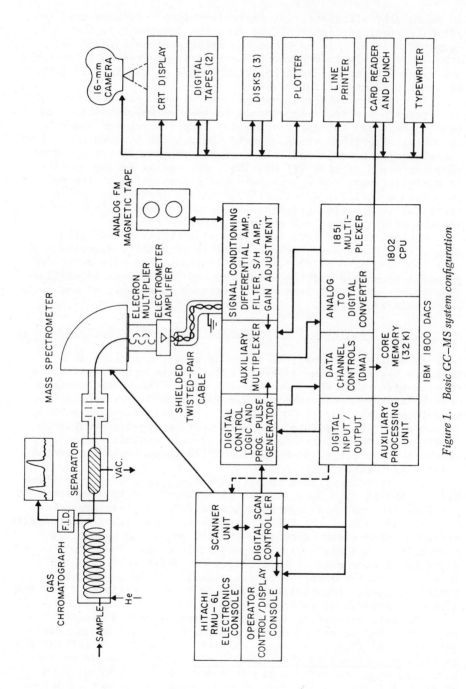

Figure 1. Basic GC–MS system configuration

filmed (5,6). Thus, all of the information produced in the
course of the experiment is permanently available to the user
independently of the computer. This approach is very economical
and more importantly, very efficient with respect to human and
computer time and resources.

To prepare the data for the several automated interpretive
techniques available (including both library searching and
several more intelligent and specialized interpretive algorithms
for specific classes of compounds), the technique of reconstruc-
tion of the spectra was developed and is being constantly improved
in this laboratory. Since this technique has been described
previously (7), a brief review of the concept will be sufficient.

Spectral data from a GC–MS analysis is acquired continually,
and at a constant rate. Thus, the normal two dimensional view
of the data and its information content (mass and intensity) can
be expanded to three dimensions to include the time element.
Components only partially resolved by the gas chromatograph will
normally show separate maxima in the mass chromatograms characteristic of those components at the times their individual con-
centrations are greatest. An enhanced set of spectra is generated
by performing a full peak profile analysis on the entire mass
chromatogram for every m/e value and then reconstructing spectra
at each scan by assembling only those m/e values which maximize
at that scan. These reconstructed spectra are practically free
of the contributions of unresolved companion substances, tailing
fractions, column bleed, and other sources of interference. The
Mass-Resolved Gas Chromatogram (the equivalent of a Total Ionization Plot generated from this new set of reconstructed data)
is generated and the entire set of data is microfilmed. The
Mass-Resolved Gas Chromatogram is now a well-resolved indication
of the number, location, and relative intensity of each of the
components in the mixture.

The reconstructed spectra are relatively free of interferences which often make the correct identification of the various
components more difficult if not impossible. Automated identification techniques such as library searching and others such
as the algorithm for the interpretation of peptide mixtures
described below are greatly facilitated, and the resultant
reliability greatly improved.

An Interpretive Algorithm for Complex Peptide Mixtures

The determination of the amino acid sequence of poly-
peptides is of great interest and has been shown to be amenable
to analysis by gas chromatography-mass spectrometry (8). In
order to clarify the ensuing discussion of an algorithm to
automate the interpretation of oligopeptide mass spectra, the
peptide sequencing strategy used in this laboratory will be
briefly described. A polypeptide, such as Subunit I of the
sweet protein Monellin (9), is partially hydrolyzed with weak

acid or enzymes to produce a complex mixture of di- to pentapeptides. These peptides are derivatized to enhance their
volatility (10) (Figure 2), and injected into a gas chromato-
graph-mass spectrometer system which results in 200-300 mass
spectra of the 15-50 oligopeptides in the mixture. One must now
locate and identify all of these oligopeptides and finally
reassemble them to determine the sequence of the original polypeptide. Since the manual identification of all the components
of these complex mixtures is a time-consuming and difficult task,
an automated procedure was needed.

Searching the unknown spectra against a library of standard
spectra is not feasible since a library of all possible di- to
pentapeptides would contain approximately 3.4 million spectra.
An interpretive algorithm is feasible, however, since the polyaminoalcohol derivatives (Figure 2) used in this laboratory
display predictable fragmentation patterns. By examining a
large number of standard peptide spectra we have found that dipeptides always exhibit an intense Z1 ion; tri- to pentapeptides
always show a prominent A2 ion, and tetra- and pentapeptides
always exhibit an intense A3 ion. In addition to these fragmentation rules, the algorithm makes use of the amino acid
composition of the original polypeptide, and the retention index
for each scan which is calculated by a previously described
method (11). Also, as has been previously shown, (12) we can
calculate the expected retention index for any oligopeptide which
is a function of its composition, but relatively independent of
the amino acid sequence.

Based on this information the algorithm shown in Table I
was developed. It should be noted that for the entries in the
peptide list in steps 1 and 3, the order of the amino acids is
not significant. Thus, all tetrapeptides which have the
composition (A,B2,C) will be represented by a single entry. In
steps 4-7, however, the order of the amino acids in specific, so
that these steps refer to sequences and not just combinations
of amino acids. Steps 1 and 3 are filters based on the amino
acid composition of the original polypeptide and the retention
index of the unknown spectra. Step 4 is a filter based on the
reliable A2 ion (Z1 for dipeptides). Similarly, step 6 is a
filter based on the known presence of an A3 ion in the spectra of
tetra- and pentapeptides. For the hypothetical example shown
in Table I an A2 ion is assumed to have been found for the partial
sequences AB.. and BA.., and an A3 ion is assumed to have been
found only for the sequences ABBC and BABC. The calculation in
step 7 is complex and will be presented in detail elsewhere (13).
It should be noted that correct identification of an unknown
spectrum is not dependent on the presence of a molecular ion, nor
any specific sequence ion (except the reliable A2, A3, and Z1 as
described above).

This algorithm was tested on the data from the analysis of a
mixture of oligopeptides generated in an acid hydrolysis experi-

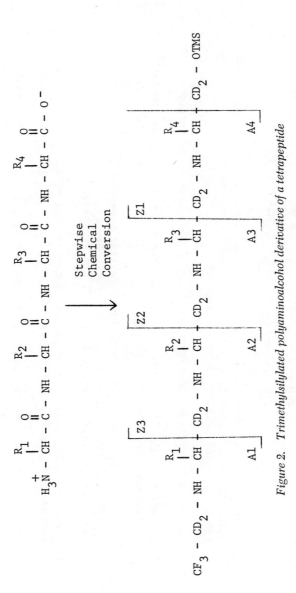

Figure 2. Trimethylsilylated polyaminoalcohol derivative of a tetrapeptide

TABLE I

PEPTIDE INTERPRETIVE ALGORITHM

1. Generate a list of all di- to pentapeptides with amino acid composition compatible with sample, and calculate their retention indices. (e.g. (A,B_2,C); R.I. $= 630 + a + 2b = c$)

2. Enter mass spectrum and retention index of next scan to be interpreted.

3. Select from the list of peptides those with retention indices similar to unknown's retention index ($\pm$ 7%).

4. For each peptide generate N-terminal partial sequences (e.g. AB.., BA.., AC.., CA.., BC.., CB.., BB..) and check for the expected A2 ion (Z1 for dipeptides).

5. Generate full sequences for partial sequences for which the A2 ion is present (e.g. ABBC, ABCB, BABC, BACB).

6. Check tetra- and pentapeptide sequences for A3 ion (e.g. ABBC, BABC).

7. Calculate scores, save best finds, and print results.

ment. Eighteen peptides had been previously identified by an
experienced human interpreter. The algorithm was able to
correctly identify seventeen of the oligopeptides. The peptide
not correctly identified was found as the third most likely
choice following two similar structures.

We have attempted in the preceding discussion to illustrate
why it is necessary to develop all aspects of a balanced data
acquisition, processing, display, and interpretive system in
order to cope with the problems inherent in the analysis of
complex mixtures by GC–MS. Although good interpretive algorithms
are very important in attempts at automated structure elucidation,
at least in the case of mixture analysis, they are at best only
as good as the weakest link in the procedure which provides the
data to be analyzed.

Literature Cited.
1. Biemann, K., Oro, J., Toulmin III, P., Orgel, L.E., Nier, A.O.,
   Anderson, D.M., Simmonds, P.G., Flory, D., Diaz, A.V.,
   Rushneck, D.R., Biller, J.E., Science (1976), 194, 72–76.
2. Watson, J.T., and Biemann, K., Analytical Chemistry (1964),
   36, 1135.
3. Biller, J.E., Ph.D. Thesis, Massachusetts Institute of Technology, May 1972, and other unpublished work.
4. Hertz, H.S., Hites, R.A., and Biemann, K., Analytical
   Chemistry (1971), 43, 681.
5. Biller, J.E., Hertz, H.S., and Biemann, K., Paper F2,
   presented at the Nineteenth Annual Conference on Mass
   Spectrometry and Allied Topics, Atlanta, Georgia, May 1971.
6. Biller, J.E., Hertz, H.S., and Biemann, K., In preparation.
7. Biller, J.E., and Biemann, K., Analytical Letters (1974),
   7(7), 518–528.
8. Biemann, K., in "Biomedical Applications of Mass Spectrometry",
   G. Waller (ed.) (1972), 405–428.
9. Hudson, G., and Biemann, K., Biochemical and Biophysical Res.
   Communications (1976), 71(1), 212–220.
10. Kelley, J.A., Nau, H., Forster, H.-J., and Biemann, K., (1975),
    Biomedical Mass Spectrometry, 2, 313–325.
11. Nau, H., and Biemann, K., Analytical Chemistry (1974),
    46(3), 426–434.
12. Nau, H., and Biemann, K., Analytical Biochemistry (1976),
    73, 139–153.
13. Herlihy, W.C., and Biemann, K., In preparation.

3

The NIH–EPA Chemical Information System

G. W. A. MILNE

National Institutes of Health, Bethesda, MD 20014

S. R. HELLER

Environmental Protection Agency, Washington, DC 20460

The quantity of data associated with analytical chemistry has been expanding very rapidly during the last twenty years or so, but until recently, the efficient application of computers to this problem has been vitiated by the high costs of computer storage and computation. With the continuing improvement in computer technology and the steady decrease in computation costs, it has, in the last two years, become feasible to consider the development of a completely interactive chemical information system.

Earlier searching systems avoided the cost of bulk storage by maintaining the data files on magnetic tape rather than disk. Tape is a very cheap form of storage but its use implies batch searching which is necessarily slow because tape is not suscept- ible to random access. Magnetic disks, on the other hand, are random access devices and the data stored on them can be searched very rapidly. Until recently, the cost of storage on disks has precluded their use for large data bases, but now it is becoming practical to consider this approach. Since this permits inter- active computing, we have developed a chemical information system that uses disk exclusively for the storage of data.

Interactive computing is a signficantly different process from batch, or off-line, computing and a different philosophy can be used in the design of programs for such work. A major problem that a chemist has in searching a chemical data base, is that the best questions to ask are often not known. An interactive system can provide the answer to a question immediately and this will enable the user to see the deficiences in the question and to frame a new query. In this way, there can be built a feedback loop in which the chemist acts as a transducer, "tuning" the query until the system reports precisely what is required.

The NIH–EPA Chemical Information System (CIS), which is described in this chapter, has been designed around this general approach (1).

System Design

The CIS consists of a collection of chemical data bases together with a battery of programs for interactive searching through these disk-stored data bases. In addition, there are a series of programs for the analysis of data, either to reduce them to a form suitable for searching purposes or as an end in itself.

The data bases that are in the CIS include files of mass spectra, carbon-13 nuclear magnetic resonance spectra, x-ray diffraction data for crystals and powders, and several biblio-graphic data bases. The analytical programs include a family of statistical analysis and mathematical modelling algorithms and programs for the calculation of isotopic enrichment from mass spectral data, analysis of nmr spectra and energy minimization of conformational structures.

a. <u>Addition of components to the CIS</u>. A general protocol for updating of CIS components or the addition to the CIS of new components has been established and a schematic diagram of this is shown in Figure 1.

In the first phase, a data base is acquired, if necessary, from one of a variety of sources. Some of the CIS data bases have been developed specifically for the CIS, an example of this being the mass spectral data base (2). Other data bases, such as the Cambridge Crystal File (3), are leased for use in the CIS and still others, such as the X-ray powder diffraction file (4), are operated within the CIS by their owners. Next, the necessary program development is undertaken. If the component is one involving searching of a data base, some reformatting of the data base, sorting and inversion of files and so on, is usually required, and this is carried out on the NIH IBM 360-168, which is well-suited to processing of large files of data. Once searchable files have been prepared, they are transferred to the NIH PDP-10 computer which is primarily a time-sharing computer, and the programs for searching through the data bases are written. The analytical, data base-independent programs of the CIS are usually written entirely on the PDP-10. Out of this work there finally emerges a pilot version of the component.

The pilot version is then allowed to run on the NIH PDP-10 and access to it is provided to a small number of people who can log into the NIH computer by telephone, using long distance calls if necessary. These users are provided with free computation and in return, they test the component thoroughly for errors and deficiencies. Such problems are reported to the development team, which attempts to deal with them. Depending upon the size and complexity of the component, this testing phase can last as long as eighteen months.

When testing is complete, the entire component is exported to a networked PDP-10 in the private sector and the version on

the NIH computer is dispensed with. The component in the private
sector is available to the general scientific community and can be
used on a fee-for-service basis. In this phase, the government
retains no financial interest in the component; it is "managed" by
a sponsor outside the U.S. Government. The Department of Industry
of the British Government, for example, maintains the Mass Spectral
Search System on the network. Advice and consultation between such
sponsors and NIH/EPA personnel continues, but the U.S. Government
does not subsidize the routine operation of CIS components in the
private sector. In fact, various government agencies of the
governmnt are actually users of the CIS and they pay according to
their use of the system, like any other user. Charges for use of
these components must be designed to cover costs, and if the
component attracts insufficient use at these prices, then it is
probably not viable and its sponsor need not continue to support
it.

     b. Computers facilities used by the CIS. Programs of the
CIS have usually been designed for use with a DEC PDP-10 computer
system. The reason for this is that the PDP-10 is one of the
better time-sharing systems available and has been adopted by a
number of commercial computer network companies as the main vehicle
for their networks. Transfer of a program from the NIH PDP-10 to
a network PDP-10 is usually straightforward, and use of a networked
computer is favored because the alternative philosophy of exporting
programs and data bases to locally operated PDP-10 computers is
less workable. This latter approach contains a number of deficiencies that are overcome by a network. Most important in this
connection is the fact that use of a networked machine means that
data bases need only be stored once, at the center of the network.
A great deal of money is thus saved because duplicate storage is
not necessary. Further, a single copy of a data base is easy to
maintain, whereas updating of a data base that resides on many
computers is virtually impossible. Finally, communications
between systems, personnel, and users is very simple in a network
environment, as is monitoring of system performance.
     For these and other reasons, the policy of disseminating the
CIS via networked PDP-10 computers was adopted at the outset and
has proved to be quite successful. A typical U.S. network of this
sort has something under 100 nodes - i.e., local telephone call
access is available in about 100 locations. These are mainly in
the U.S., but a substantial number will be found in Europe.
Further, some computer networks are now themselves interfaced to
the Telex network, thus making their computer systems available
worldwide. Irrespective of one's location, the cost of access is
somewhere between $7 and $15 per hour, depending upon the trans-
mission speed used and also on the time of day. Networks usually
offer 110, 300 and 1200 baud service and the response time of the
system is usually negligibly small.

The only equipment that is required to establish access to a computer network, is a telephone-coupled computer terminal. Typewriter terminals are becoming very common and are also becoming relatively cheap. Such a terminal can be purchased from a variety of manufacturers for between $1,000 and $3,000 and in general, will operate at 300 baud (30 characters/second). A cathode ray terminal, capable of running at 1200 baud can be purchased for as little as $2,000. Any equipment of this sort can usually be leased or purchased.

Components of the CIS

a. <u>Mass Spectral Search System</u>. The Mass Spectral Search System (MSSS) is the oldest component of the CIS, having been developed in 1971, and has been seen as a prototype for more recent components. Developed as a joint effort between NIH, EPA, and the Mass Spectrometry Data Centre (MSDC) in England, the current MSSS data base contains about 30,000 mass spectra representing the same number of compounds. This has been derived from an archival file containing some 60,000 spectra of the same 30,000 compounds(5). Computer techniques have been employed to assign every spectrum a quality index(6) and where duplicate spectra appear in the archive file, only the best spectrum is used in the working file. All compounds in the archive have been assigned Chemical Abstracts Service (CAS) registry number, a unique identifier that is used to locate duplicate entries for the compounds, find the compound in other CIS files and provide structure and synonym lookup capabilities throughout the CIS.

Searches through the MSSS data base can be carried out in a number of ways. With the mass spectrum of an unknown in hand, the search can be conducted interactively, as is shown in Figure 2. In this search the user finds that 24 data base spectra have a base peak (minimum intensity 100%, maximum intensity 100%) at an m/e value of 344. When this subset is examined for spectra containing a peak at m/e 326 with intensity of less than 10%, only 2 spectra are found. If necessary, the search can be continued in this way until a manageable number of spectra are retrieved as fulfilling all the criteria that the user cited. These answers can then be listed as is shown in Figure 2. Alternatively, the file can be examined for all occurrences of a specific molecular weight or a partial or complete molecular formula. Combinations of these properties can also be used in searches. Thus, all compounds containing, for example, five chlorines and whose mass spectra have a base peak at a particular m/e value can be identified.

In contrast to these interactive searches, which are of little appeal to those with large numbers of searches to carry out, there is available a batch-type search which accepts the complete spectrum of the unknown and sequentially examines all spectra in the file to find the best fits. A user's data system can be connected to the network for this purpose and the unknown

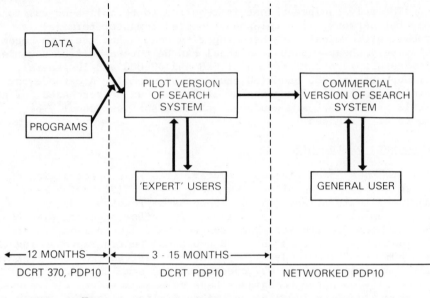

Figure 1.   Protocol for adding a component to the CIS

```
USER RESPONSE:PEAK

TYPE PEAK,MIN INT,MAX INT
CR TO EXIT,'1 FOR ID#,REGN,QI,MW,MF AND NAME

USER:344,100,100
  # REFS   M/E PEAKS

        24         344

NEXT REQUEST: 326,0,10
  # REFS   M/E PEAKS

        2          344 326

NEXT REQUEST: 1
   ID#            REGN  QI MW   MF                              NAME

23455           19594913 624 344 C22H3203              A-Nor-5.alpha.-
                                             androstan-3-one, 17.beta.-hydr
                                             oxy-5-vinyl-, acetate (8CI)
25330           18326153 974 386 C24H3404              Podocarpa-8,11,
                                             13-triene-3.beta.,12-diol, 13-
                                             isopropyl-, diacetate (8CI)
```

Figure 2.   PEAK search in the MSSS

spectra can be down-loaded into the network computer for use in
this search, which can be carried out at once, or, if preferred,
overnight at 30% of the cost.

Once an identification has been made, the name and registry
number of the data base compound are reported to the user. If
necessary, the data base spectrum can be listed or, if a CRT
terminal is being used, plotted, to facilitate direct comparison
of the unknown and standard spectra.

The MSSS has been generally available through computer networks for several years and is now currently resident upon the
ADP-Cyphernetics network where some 3,000 searches and 2,000 other
transactions, such as retrievals, are carried out each month by
the approximately 200 users. All searches in the MSSS are transaction priced at between $1 and $7 and in addition to these
charges and the connect time charge, users must pay the annual
subscription fee of $300. This fee is used to defray the annual
disk storage charges which are paid in advance by the sponsor of
the MSSS, the Department of Industry of the British Government.

b. Carbon Nuclear Magnetic Resonance (CNMR) Spectral Search
System. The data base that is used in the CNMR search system
consists currently of 4,400 CNMR spectra. As in the case of the
MSSS, every compound has a CAS registry number, and all exact
duplicates have been removed from the file. A specific compound
may still appear in this file more than once, however, because its
CNMR spectrum may have been recorded in different solvents. The
CNMR file is still small but is growing at a fairly steady rate
and should benefit considerably from recent international agree-
ments to the effect that all major compilations of CNMR data will,
in the future, be pooled.

Searching through this data base, as in the case of the MSSS,
can be interactive or not. In the interactive search, a user
enters a shift, with an acceptable deviation, and the single
frequency offresonance decoupled multiplicity, if that is known.
The program reports the number of file spectra fitting one or both
of these criteria. The names of the compounds whose spectra have
been retrieved can be listed, or alternatively, the list can be
reduced by the entry of a second chemical shift. A search for
spectra of compounds having a specific complete or partial molecular formula can also be carried out, but there is no capability
for searching on molecular weight, a parameter of little relevance
to CNMR spectroscopy.

If an interactive search is not appropriate to the problem at
hand, a batch type of search through the data base using the
techniques described by Clerc et al.(7) is available. To carry
out such a search, the user enters all the chemical shifts from
the unknown and starts the search. The entire unknown spectrum is
compared to every entry in the file and the best fits are noted
and reported to the user. This program searches for the absence
of peaks in a given region as well as for the presence of peaks

and thus has the capability of finding those compounds which are structurally similar to the material that gave the unknown spectrum.

When a search is completed, the user is provided with the accession numbers of spectra that fit the input data. The names and CAS registry numbers of the compounds in question will also be given. If more information is required, the complete entry for a given accession number can be retrieved. This includes a numbered structural formula, the name, molecular formula and registry number of the compound, experimental data pertaining to the spectrum and the entire spectrum, together with single frequency off-resonance decoupled multiplicities and (if available) relative line intensities and assignments.

This CNMR search system recently has been made available on the ADP-Cyphernetics network. Searches are all transaction priced at $1-3.

c. X-ray Crystallographic Search System. This is a series of search programs working against the Cambridge Crystal File(8), a data base of some 15,000 entries dealing with published crystallographic data mainly for organic compounds. The entry for each compound contains the compound name, its molecular weight and registry number, the space group in which it crystallizes and the parameters of the unit cell of the crystal. The file may be searched on the basis of any of these parameters as shown in Figure 3, which is an example of a search for any compounds that crystallize in space group P 1 and have a molecular weight between 250 and 300. As can be seen from Figure 3, there are 98 entries with the correct space group (scratch file 1) and 867 with a molecular weight within the specified range (scratch file 3). The intersection of these files reveals that only 3 compounds (scratch file 4) meet both specifications, and the first of these compounds, crystal sequence number 4413, is listed.

All the compounds in this file have been registered by the CAS and their connection tables have been merged into the file. This data base is, therefore, searchable on a structural or substructural basis, as are all the othe files of the CIS.

Once an entry of interest in this file has been located by one of the search programs, its file accession number, the "crystal sequency number" can be used to retrieve the appropriate literature reference or the structure, or both.

This file is available for general use via the ADP-Cyphernetics network. Currently, the charging of options in this system is not transaction-priced. Enough statistics are now available to indicate that all searches, other than structural searches, cost less than $2.00 and that the structural searches cost possibly as much as $10.00.

d. X-ray Crystal Data Retrieval System. The National Bureau of Standards (NBS) has collected a file of data pertaining to some

```
*SPGR
SPACE GROUP SYMBOL
> P 1
FILE = 1 REFERENCES =   98 ITEM = P 1
*SMOLS 1
TYPE MOLECULE WEIGHT RANGE

>250,300
FILE = 3   MERGED REFERENCES = 867

*INTER 1 3
FILE = 4    INTERSECTED REFERENCES =    3
SOURCE FILES WERE:   1   3
*SSHOW 4
START WITH NTH REFERENCE   (1)=1
SHOW EVERY NTH REFERENCE   (1) =1
STRUCTURE     1 CRYSTAL SEQUENCE      4413
```

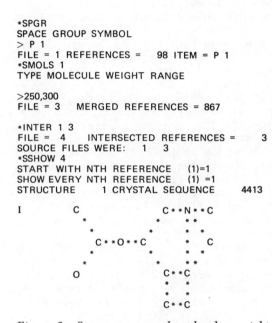

*Figure 3. Space group and molecular weight
search in the Cambridge crystal file*

24,000 crystals, including those in the Cambridge file described
in (c) above (9). The data in the NBS file include the cell
parameters, the number of molecules, Z, in the unit cell, the
measured and calculated densities of the crystal and two determin-
ative ratios, such as A/B and A/C. Every compound in the file is
identified by its name, molecular formula, and registry number,
and the file can be structurally searched by the CIS substructure
search system which is described below.

Searches through this data base for crystals with specific
space groups, or densities are possible and crystals with unit
cells of given dimensions can also be found. It is hoped that
this may prove to be a very rapid method of identifying compounds
from the readily measured crystal properties.

e. <u>X-ray Powder Diffraction Retrieval Program</u>. Compounds
that fail to crystallize may still be examined by X-ray diffraction, because powders give characteristic diffraction patterns. A
collection of powder diffraction patterns proves to be a very
effective means by which to identify materials and indeed, one of
the very earliest search systems in chemical analysis was based
upon such data by Hanawalt (10) nearly forty years ago. The data
base of some 27,000 powder diffraction patterns(11) that is used
in the CIS is in fact a direct descendant of that with which
Hanawalt carried out his pioneering work. A problem that arises
in connection with this particular component stems from the fact
that powders, as opposed to crystals, are frequently impure. The
patterns that are obtained experimentally, therefore, are often
combinations of one or more file entries. A reverse searching
program, that examines the experimental data to see if each entry
from the file is contained in it(12), has been written and seems
to cope with this particular difficulty. It is currently running
in test on the NIH PDP-10 and will be made available to the
scientific community during 1977.

f. <u>Substructure Search System (SSS)</u>. All the compounds in
the files of the CIS have been assigned a registry number by the
CAS. The registry number is a unique identifier for that compound,
and may be used to retrieve from the CAS Master Registry, all the
synonyms that the CAS has identified for the compound, these being
names that have been used for the compound, in addition to the
name used in the CAS 9th Collective Index. Further, the registry
number can be used to locate in the CAS files, the connection
table for the compound's structure. This is a two-dimensional
record of all atoms in the molecule together with the atoms to
which each is bonded and the nature of the bonds (13). This
connection table is the basis of the substructure search component
of the CIS (14).

The purpose of the SSS is to permit a search for a user-
defined structure or substructure through data bases of the CIS.
If a substructure is found to be in a CIS data base, then armed

with its registry number, the user can access that file and locate the compound and hence inspect whatever data is available for it.

As the first step in this process, the user must, of course, be able to define the structure of interest to the computer. This is done with a family of structure generation programs which can, for example, create a ring of a given size, a chain of a given length, a fused ring system and so on. Branches, bonds and atoms can be added and the nature of bonds can be specified. The element represented by any nodes can be defined; in the absence of such definition, the atom is presumed to be carbon. As the query structure is developed using these commands, the computer stores the growing connection table. If the user wishes to view the current structure at any point, the display command (D) can be invoked. This command, using the current connection table, generates a structure diagram that can be printed at a conventional terminal.

When the appropriate query structure has been generated, a number of search options can be invoked to find occurrences of this query structure in the data base. The two most useful search options are the fragment probe and the ring probe. The fragment probe will search through the assembled connection tables of the data base for all occurrences of a particular fragment, i.e., a specific atom, together with all its neighbors and bonds. The user may specify particular fragments in the query structure which are thought to be fairly unique and characteristic of that structure. Alternatively, a search for every fragment in the query structure may be requested. The general form of a fragment probe is as shown in Figure 4. The query structure contains only one relatively unique node, C3, and this is the one which is sought in the data base. It is found to occur 980 times and a temporary file of just those particular entries is stored as file #2. This can be accessed by the user either for the purpose of listing its contents, as is shown in the figure, or to intersect it with other scratch files.

The ring probe search is a search through the data base for all structures containing the same ring or rings as the query structure. A ring that is considered to be an answer to such a query must be the same size as that in the query structure. It must also contain the same number of non-carbon atoms (heteroatoms) but the nature of the heteroatoms and the position of any substituents can be required by the user to be the same or different to that in the query structure. Thus with a query structure of pyrrole, the only "exact" answer is pyrrole but the user may permit the retrieval of "imbedded" answers which would include furan and thiophene. Similarly, o-xylene itself is the only "exact" match for o-xylene, but m- and p-xylene would be considered as "imbedded" matches. An example of a ring probe search is given in Figure 5. Here the query structure is a 3,4-dichlorofuran, but imbedded matches for heteroatoms and substitutents have been

OPTION? D

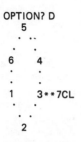

OPTION? FPROB 3

TYPE E TO EXIT FROM ALL SEARCHES,
T TO PROCEED TO NEXT FRAGMENT SEARCH

FRAGMENT:

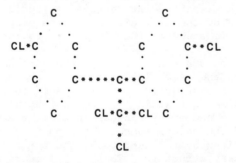

REQUIRED OCCURRANCES FOR HIT : 1
THIS FRAGMENT OCCURS IN   980 COMPOUNDS

FILE = 2,   980 COMPOUNDS CONTAIN THIS FRAGMENT

OPTION? SSHOW 2
HOW MANY STRUCTURES (E TO EXIT) ? 1
STRUCTURE   1 CAS REGISTRY NUMBER   50293

C14H9Cl5

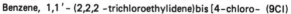

Benzene, 1,1' - (2,2,2 -trichloroethylidene)bis [4-chloro- (9CI)

Figure 4.   Fragment probe in the substructure search system

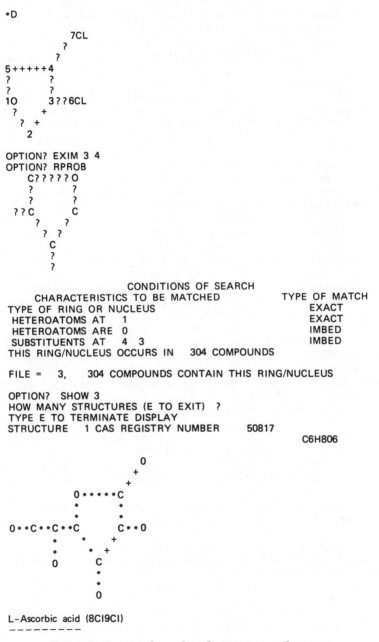

```
*D

                 7CL
             ?
           ?
5+++++4
?        ?
?        ?
1O       3??6CL
  ?    +
    ? +
      2
```

OPTION? EXIM 3 4
OPTION? RPROB
```
     C?????O
     ?         ?
     ?         ?
   ??C         C
      ?     ?
       ? ?
         C
         ?
         ?
```

                    CONDITIONS OF SEARCH
        CHARACTERISTICS TO BE MATCHED   TYPE OF MATCH
TYPE OF RING OR NUCLEUS   EXACT
  HETEROATOMS AT   1   EXACT
  HETEROATOMS ARE   0   IMBED
  SUBSTITUENTS AT   4 3   IMBED
THIS RING/NUCLEUS OCCURS IN   304 COMPOUNDS

FILE =   3,   304 COMPOUNDS CONTAIN THIS RING/NUCLEUS

OPTION? SHOW 3
HOW MANY STRUCTURES (E TO EXIT) ?
TYPE E TO TERMINATE DISPLAY
STRUCTURE   1 CAS REGISTRY NUMBER   50817
                                             C6H8O6
```
          0
          +
          +
   0 * * * * *C
   *         *
   *         *
0 * *C * *C * *C        C * *0
   *       *     +
   *       *   +
   0       C
           *
           *
           0
```

L–Ascorbic acid (8CI9CI)
– – – – – – – – –

Figure 5.   Ring probe in the substructure search system

allowed and so the list of 304 answers will include any disubstituted pyrrole as well as any disubstituted furan and so on.
Higher substitution will also be permitted.

In addition to these structural searches, there are a number
of "special properties" searches that often prove to be very useful as a means of reducing a large list of answers resulting from
structure searches. The special properties searches include
searches for a specific molecular weight or range of molecular
weights and a search for compounds containing a given number of
rings. Searches may also be conducted for the molecular formula
corresponding to the query structure, or for a different user-
defined molecular formula. This may be specified completely or
partially and the number of atoms of any element may be entered
exactly or as a permissible range.

If one's purpose is to determine only the presence or absence
in a data base of a specific structure, this can be accomplished
with the search option "IDENT". This
program hash-encodes the query structure connection table and
searches through a file of hash-encoded connection table for an
exact match. The search, which is very fast by substructure
search standards, has been designed specifically for those users
who, to comply with the Toxic Substances Control Act(15), have to
determine the presence or absence of specific compounds in
Environmental Protection Agency files.

Finally, if one has completed ring probe and fragment probe
searches for a specific query structure and is still confronted
with a sizeable file of compounds that satisfy the criteria that
were nominated, a substructure search through this file may be
carried out. This involves an atom-by-atom, bond-by-bond comparison of every structure.

The substructure search system is currently operating on 19
files which are given in Table 1. The whole system is available
for general use on the Tymshare computer network. A subscription
fee of $150 per year must be paid for use of the system and the
only other charges are the connect-time charge and the searching
costs which range upwards from $3.00-5.00 for an identity search.

g. Mass Spectrometry Literature Search System. The
accumulated files of the Mass Spectrometry Bulletin, a serial
publication of the Mass Spectrometry Data Centre, Aldermaston,
England, have been made the basis of an on-line search system.

The Bulletin, which since 1967, has collected about 60,000
citations to papers on mass spectrometry, may be searched interactively for all papers by given authors, all papers dealing with
one or more specific subjects or with one or more particular
elements. In addition, citations dealing with general index
terms may also be retrieved. Simple Boolean logic is available,
and thus searches may be conducted for papers by Smith and Jones,
or Smith but not Jones, and so on. Citations retrieved may be
limited to specific publication years, between 1967 and the

TABLE 1. COMPONENT FILES OF THE NIH-EPA
SUBSTRUCTURE SEARCH SYSTEM.

Cambridge (Xray) Crystal File.
CPSC Chemicals in Consumer Products.
EPA AEROS SOTDAT File.
EPA Las Vegas Chemical Spill File.
EPA Storage and Retrieval of Air Data.
EPA Pesticide Standards.
EPA STORET Water Data Base.
EPA-FDA Pesticide Repository Standards.
EPA Inactive Ingradients in Pesticides.
EPA Oil and Hazardous Materials File.
EPA Pollutants in Drinking Water.
EPA Pesticides File.
EROICA Thermodynamics Data File.
Merck Index.
NBS Gas Phase Proton Affinities.
NBS Heats of Formation of Gaseous Ions.
NBS Single Crystal File.
NCI-SRI Industrial Chemicals File.
NCI PHS-149 File of Carcinogens.
NIMH File of Psychotropic Drugs.
NIH-EPA Carbon-13 Nuclear Magnetic Resonance Search System.
NIH-EPA Mass Spectral Search System.
WHO International Non-proprietary Name File of Drugs.

present. The interactive nature of the search provides great
control to the user. One can learn within a few minutes that
while there are in the Bulletin, 463 papers dealing with mass
measurement for example, and 678 on chemical ionization, only 8
report on mass measurement in chemical ionization mass spectra.
Similarly, one can rapidly discover that although there are 532
papers dealing with carbon dioxide, only 1 of these was presented
at the 1975 NATO meeting in Biarritz.

No numerical codes are used by the system. A search for a
specific subject can be carried out by entering the subject word
itself. If the word "mass" is entered, searches for 7 terms (all
those containing the fragment "mas", i.e., mass spectra, mass
discrimination, mass measurement and so on) are conducted and the
user is asked to select the one of interest. In this way,
knowledge of the correct subject words or of their correct
spelling is not necessary.

The whole search system has been written for use on a PDP-10
computer and is a component of the MSDC-NIH-EPA Mass Spectral
Search System. As such, it is accessible via the ADP-Cyphernetics
computer network.

h. <u>X-ray Crystal Literature Retrieval System</u>. The data
base used in the X-ray crystallographic search system described
in (c) above possesses complete literature references to all
entries in the file (<u>8</u>). This information has been made the
basis of a system for searching the literature pertaining to the
X-ray diffraction study of organic molecules.

As in the Mass Spectrometry Bulletin Search System, it is
possible to search for papers by a specific author(s), and
papers that appeared in given years in given journals may also be
retrieved. Additionally, papers may be located on the basis of
specific words appearing in their titles. These words may be
truncated by the user and so the fragment "ERO" will retrieve
papers with the word "STEROID" on their titles or papers whose
titles use the word "MEROQUININE". The system generates scratch
files from searches, as in the substructure search system, and
files can be intersected upon request with "AND" or "NOT" operators.
Thus one could, for example, retrieve all papers published in
Acta Crystallographica since 1970 by Atkins, excluding specifically those on corticosteroids.

Once a paper of interest has been identified, all the
crystallographic information in that paper can be examined
because the crystal serial number of the paper can be used in the
crystallographic search system to retrieve that information.
Alternatively, the CAS registry number of any particular compound
can be used to retrieve any data of interest on that compound
from other files of the CIS.

The X-ray literature search system is operating on the ADP-
Cyphernetics network as a component of the X-ray crystallographic
search system. Searches are not transaction priced but cost
under $2.00 on average.

i. <u>Proton Affinity Retrieval Program</u>. With the current
high level of interest in chemical ionization mass spectrometry,
there is a need for a reliable file of gas phase proton affinities. No data base of this sort has previously been assembled
and for these reasons, the task of gathering and evaluating all
published gas phase proton affinities has been undertaken by
Rosenstock and coworkers at NBS. This file (16), which has about
400 critically evaluated gas phase proton affinities drawn from
the open literature, can be searched on the basis of compound
type or of the proton affinity value. It will be appended to the
MSSS and the bibliographic component will be merged with the Mass
Spectrometry Bulletin Search System.

j. <u>NMR Spectrum Analysis Program</u>. Many proton nmr spectra
can be satisfactorily analyzed by hand, and such first order
analysis is, in these cases, a quite satisfactory way of assigning
chemical shifts and coupling constants to the various nuclei
involved. In certain cases, however, so-called second order
effects become important and as a result, more or fewer spectral
lines than are indicated by first order considerations will
result. A way to analyze such spectra is to estimate the various
coupling constants and chemical shifts and then, using any of a
variety of standard computer programs (17), calculate the
theoretical spectrum corresponding to these values. The calculated
spectrum can be compared to the observed spectrum and a new
estimate of the data can be made. In this way, by a series of
successive approximations, the correct coupling constants and
chemical shifts can be determined.

The CIS component GINA (Graphical Interactive NMR Analysis)
which is based upon the programs developed by Johannesen et
al.(18), permits these operations in real time in an interactive
fashion. The program is designed for use with a vector cathode
ray tube terminal upon which each new theoretical spectrum can be
displayed for comparison by the user with the observed spectrum.
The program has been available at NIH for over four years (19)
and is currently being exported to a computer network in the
private sector. The cost of using the program is not yet well
established because it is subject to wide variations.

k. <u>Mathematical Modelling System(MLAB)</u>. MLAB is a program
set, developed by Knott at NIH (20) which can assimilate a file
of experimental data, such as a titration curve and perform on it
any of a wide variety of mathematical operations. Included
amongst these are differential and integral calculus, statistical
analysis (mean and standard deviation, curve and distribution
fitting and linear and non-linear regression analysis). Output
data can be presented in any form, but the PDP-10-resident
program is especially powerful in the area of graphical output.

Data can be displayed in the form of two- or three-dimensional
plots which can be viewed and modified on a CRT terminal prior to
pen-and-ink plotting.

This program set is now available for general use on the
ADP-Cyphernetics network. The cost of using MLAB is based upon
the computer resource units and depends, therefore, upon the type
of work that is being done.

1. Isotopic Label Incorporation Determination (LABDET).
Radioisotopes are particularly well-suited to labelling studies
because they can be very easily detected at very low levels. In
recent years, however, there has been increasing concern about
the shortcomings of radioisotopes in medical research. Current
standards, in fact, take the position that the use of radioisotopes such as carbon-14 in children and women of child-bearing
age is precluded. Consequently, it is not possible to study the
metabolism of drugs in such patients using radioisotopes, and
this leads to some difficulty because it is only in such patient
groups that the metabolism of drugs, such as oral contraceptives,
is of relevance.

Much effort has gone into studies of the possibilities of
carbon-13 as a surrogate for carbon-14, and this type of work
applies also to problems involving oxygen and nitrogen which have
stable isotopes, but no convenient radioisotopes. Mass spectrometry is the best general method for detection and quantitation of
stable isotopes in molecules, but there a serious problem involved
in its application is that stable isotopes such as carbon-13,
nitrogen-15 and oxygen-18 occur naturally as minor components of
natural elements. This is most pronounced in the case of carbon.
Naturally-occurring carbon is about 99% C-12 but a small variable
amount of all natural carbon is C-13. This creates a "background" against which determinations of isotope levels in
labelled compounds must be measured. The purpose of LABDET(21)
is to compare the mass spectrum of an unlabelled compound with
that of the same compound isotopically enriched. This is usually
done using the molecular ion region of the spectra. The program
calculates an estimate of the level of incorporation of isotope
and then calculates a theoretical spectrum which can be compared
with the actual spectrum. The theoretical spectrum is then
adjusted and a further comparison is made, and in this way, the
program proceeds through a predetermined number of iterations,
finally calculating the correlation coefficient between the
observed spectrum and the best theoretical spectrum.

This calculation is not difficult so much as tedious and if
one must carry it out many times per day, use of the computer is
indicated. LABDET is an option within MSSS on the ADP-Cyphernetics network and its use costs $2.00.

m. Conformational Analysis of Molecules in Solution. A
problem of long standing in chemistry has been to estimate the

relationship between the conformation of a molecule in the crystal, as measured by X-ray methods, with that in solution where barriers to rotation are greatly reduced. A sophisticated program set for Conformation Analysis of Molecules in Solution by Empirical and Quantum-mechanical methods (CAMSEQ) has been developed for this purpose by Hopfinger and coworkers (22) at Case Western Reserve University.

This program can run in batch or interactively. As input data, it requires the structure of the compound and this can be provided as a set of coordinate data from X-ray measurements, it can be entered interactively in the form of a connection table or the program can simply be provided with a CAS registry number, and if the corresponding connection table is in the files of the CIS, it will use that.

The first task is to generate the coordinate data corresponding to a particular compound. Then the free energy of this conformation in solution is calculated, Next the program begins to change torsion angles specified by the user in the conformation and with each new conformation, a statistical thermodynamic probability is calculated, based upon potential (steric, electrostatic, and torsional) functions and terms for the free energy associated with hydrogen-bonding, molecule-solvent, and molecule-dipole interactions. This program, in its interactive version, can be run in under 40K words of core. and work is in progress to export it to a commercial networked computer.

Conclusions

One of the first goals of the CIS was to produce a series of searchable chemical data bases for use by working analytical chemists with no especial computer expertise. A second aim was to link these data bases together so that the user need not be restricted to a consideration of, for example, only mass spectral data.

The various problems inherent in these plans included acquisition of data bases, design of programs, dissemination of the resulting system and linking, via CAS registration numbers, of the various CIS components. These problems, as has been described above, have been solved conceptually and, to a large extent, practically, and the CIS, as it now stands, is the result. It is now possible, therefore, to review the system in an effort to define future goals, and a number of these seem fairly clear.

Searches through more than one data base in combination would be very desirable. For example, one often possesses mass spectral and nmr data for an unknown, and it would very useful to be able to identify any compounds that match these data in a single search. In another development, it is expected that the CONGEN program developed for the DENDRAL project (23) will be merged into CIS during the coming year. This program, which generates structures corresponding to a specific empirical formula, could be extremely useful in a strategy for structure solving using the CIS. It is not at all difficult to envisage situations in which a reduced set of structures could be produced by CONGEN for consideration. Each structure in turn could be used as an input in the substructure search system, and the various compounds whose registry numbers are so retrieved could be considered to be possible answers to the problem. Confirmation for any of them could then be sought in the spectral data bases, the registry number being all that is necessary to locate and retrieve data. One can even speculate further to the day when synthetic pathways to any likely candidates could be designed by the computer system which could easily add the very practical touch of checking that any starting materials for such syntheses are commercially available at an appropriately low cost!

In a different approach, the power of pattern recognition techniques could be assessed within some of the very large files contained in the CIS. This is a very useful exercise because there is little reported work of this sort on large files and thus we have begun to explore the value of such methods in handling the problem of identification of true unknowns such as water pollutants. Programs designed to test mass spectra for the presence of the compound of oxygen or nitrogen are currently being tested (24) and their utility as prefilters on mass spectral data prior to data base searching will be tested as soon as feasible.

In summary, it is felt that progress to date with the CIS has demonstrated economic feasibility in that a number of relatively stable CIS components have now been in the private sector for some time. The test before us is whether we can capitalize on this to explore the new and exciting possibilities that lie ahead in the area of structure determination by computer.

Literature Cited

1. Heller, S. R., Milne, G. W. A., and Feldmann, R. J. Science, (1977) 195, 253.
2. Heller, S. R., Fales, H. M., and Milne, G. W. A. Org. Mass Spectrom., (1973) 7, 107; Heller, S. R., Koniver, D. A., Fales, H. M., and Milne, G. W. A. Anal. Chem. (1974) 46, 947; Heller, S. R., Feldmann, R. J., Fales, H. M., and Milne, G. W. A. J. Chem. Soc., (1973) 13, 130; Heller, R. S., Milne, G. W. A., and Heller, S. R. J. Chem. Inf. Comp. Sci. (1976) 16, 176.
3. This data base (ref. 8) is leased by NIH on behalf of the entire U.S. scientific community.
4. McCarthy, G. and Johnson, G. G., paper C3 presented as a part of the Proceedings of the American Crystallographic Association meeting, State College, Pa., 1974.

5. Heller, S. R., Milne, G. W. A., and Feldmann, R. J. J. Chem. Inform. Comp. Sci (1976) 16, 232.
6. This is carried out using an unpublished program developed by McLafferty and co-workers at Cornell University.
7. Schwarzenbach, R., Meili, J., Koenitzer, H., and Clerc, J. T. Org. Magn. Res (1976) 8, 11.
8. Kennard, O., Watson, D. G., and Town, W. G. J. Chem. Doc. (1972) 12, 14.
9. These data are available as NBS Tape #9 through the National Technical Information Service, Springfield, VA 22151.
10. Hanawalt, J. D., Rinn, H. W., and Frevel, L. K. Ind. Eng. Chem. (Anal.) (1938) 10, 457.
11. This file is a proprietary product of the Joint Committee on Powder Diffraction Standards, 1601 Park Lane, Swarthmore, PA 19801.
12. Abramson, F. P. Anal. Chem. (1975) 47, 45.
13. Chemical Abstracts Service Standard Distribution Format File, 1976. Chemical Abstracts Service, Columbus, Ohio 43210.
14. Feldmann, R. J., Milne, G. W. A., Heller, S. R., Fein, A., Miller, J. A., and Koch, B. A. J. Chem. Inf. & Comp. Sci., (1977) in press.
15. Toxic Substances Control Act, Public Law 94-469, enacted October, 1976.
16. Hartmann, K., Lias, S., Ausloos, P. J., and Rosenstock, H. M. Publication NBSIR 76-1061, July, 1976.
17. Castellano, S. and Bothner-By, A. A. J. Chem. Phys.(1964), 41, 3863; Swalen, J. D. and Reilly, C. A. J. Chem. Phys. (1965) 42, 440.
18. Johannesen, R. B., Ferretti, J. A., and Harris, R. K. J. Magn. Res. (1970) 3, 84.
19. Heller, S. R. and Jacobson, A. E. Anal. Chem. (1972) 44, 2219.
20. Knott, G. D., and Shrager, R. I. Assn. Comp. Machin. SIGGRAPH Notes 6 (1972) 138.
21. Hammer, C. F. Department of Chemistry, Georgetown University, Washington, DC, unpublished work.
22. Weintraub, H. J. R. and Hopfinger, A. J. Intnl. J. Quant. Chem. (1975) 9, 203.
23. Carhart, R. E., Smith, D. H., Brown, H. and Djerassi, C. J. Amer. Chem. Soc. (1975) 97, 5755.
24. Meisel, W., Jolley, M. and Heller, S. R., in preparation.

4

An Information Theoretical Approach to the Determination of the Secondary Structure of Globular Proteins

JAMES A. DE HASETH and THOMAS L. ISENHOUR

Department of Chemistry 045A, University of North Carolina, Chapel Hill, NC 27514

The determination of three-dimensional structures of proteins directly from their amino acid sequences is a major problem in molecular biology. One facet of the overall three-dimensional structure, secondary structure, is in part described by alpha helical and beta sheet conformations. These conformations are difficult to determine in solution and even analysis by X-ray diffraction techniques is a long and complex task given that the appropriate crystals can be grown. For these reasons the development of a reliable method to aid in the assignment of secondary structure to proteins directly from the amino acid sequences would be useful. Several secondary structure prediction methods have been formulated (1-27) and schemes employing several of these methods (28-30) have been developed; however, none of these systems is capable of a high degree of correlation between observation and prediction.

Information theory has not previously been well exploited in the assignment of protein secondary structure. The method described here makes use of third-order peptide approximations to the structure of the proteins. Once these third-order approximations have been formulated they are used to decode the amino acid sequence of a protein and assign secondary structure to the decoded fragments. In this manner alpha helices and beta sheets may be predicted within an entire protein sequence.

Theory

Information theory may be easily described in the context of the written English language. The illustration used here follows principles stated by Shannon and Weaver (31). The first supposition of the illustration is that there are only twenty-seven symbols in the written English language, specifically, twenty-six alphabetic characters and a blank. A zero-order approximation to the language is defined as having all symbols independent and equiprobable. An example of a message constructed under these criteria is as follows:

46

PROJDQJPDNZZBUE CUXYRQVBENIJEWHDGEAIWBREGYDNQEKUZUWEKXTRJRQYWUZ-
ZBTWIJG WCUBONAOCM.

 As can be seen in this rather small sample the twenty-seven
symbols are fairly evenly distributed, as would be expected.
Each symbol has a 1/27 chance of being selected and the information content of this collection of symbols, or message, is very
low. The most the message can hope to convey is a list of all
possible symbols.
 A further criterion may be imposed on the approximation:
that the symbols are not equiprobable but have probabilities
based on their frequencies of occurrence in the written English
language. These new criteria define a first-order approximation.
An example is as follows:

FSGEHTETMO HONT I SDHEAPC MBDNNIOL AGWEONEEYSEN ODLU OEY S.

This message more closely approximates the written English
language than the zero-order approximation. The symbols E, O, T
and N occur rather frequently; others such as P, C and W occur
only once, whereas X, Z, J, K, Q and V are not represented at all
in this short message. These data are more consistent with
English and the message reflects these facts.
 Second-order approximations are constructed by selecting the
first symbol using the first-order criteria, the succeeding
symbol is then chosen by its frequency of occurrence when it is
preceded by the first symbol. The third is chosen by its depend-
ency upon the second symbol. This process is repeated until the
message is terminated. The symbols are no longer independent but
are biased by the preceding character. A second-order approximation message has been constructed:

INOMETHE COALT OROSIE O MY ARONANIF IVIVIF ENTHICHATOFATR.

As can be seen two and three letter words and word fragments are
formed and the message for the first time begins to have an
English appearance.
 When the frequencies of occurrence of symbols are biased by
the two immediately preceding symbols, a third-order approximation results. An example of a third-order approximation message
is:

CIENCEMAT WE LINEREMPHYSION ATICESESITY PHAW.

The word and word fragments are now larger than before; for example, CIENCE as in science; LINE; EMPH as in emphasis; PHYS as
in physical; etc. These processes can be continued until nth-
order approximations are formed; however, it is more instructive
at this point to jump to approximations that use larger discrete
units than the twenty-seven symbols.

A first-order word approximation message is constructed by randomly selecting words dependent upon their frequencies of occurrence in English. Here the words are independent but not equiprobable, for example:

BEFORE THAT ISSUES POTENTIAL CREATION WHETHER THE THROUGHOUT IN OF TO OF FOLLOWING HIMSELF WHEREAS FOR OCCUPATIONS.

Two and three word phrases have been formed and although the message obviously has an English origin very little information is conveyed. In a manner similar to the second-order approxima-tion a second-order word approximation may be formed. As can be seen in the following message multiword phrases result.

THE DOOR AND THEN HE THOUGHT OF THE MAN WILL PROTEST.

As the order of the word approximations increases sentences are formed and messages are conveyed.

Protein sequences are analogous to the written English language to the extent that specific amino acid sequences form proteins that possess certain biological activity. In this sense a protein represents a "message". The "symbols" of the protein "message" are atoms of carbon, hydrogen, oxygen, nitro-gen and sulfur, as well as single and double covalent bonds, and hydrogen bonds. These "symbols" form discrete units or amino acids under the constraints of protein chemistry. The amino acids are discrete units as are words in a written language; but the number of these discrete units (peptides) commonly found in globular proteins totals only twenty. As was illustrated above, a good representation of written English language was made with a second order word approximation. Because proteins have an extremely small "vocabulary" (twenty amino acids) a low order peptide approximation may represent protein messages to a high degree of accuracy. The necessary criteria for the formation of alpha helices and beta sheets in protein sequences may be deter-mined using nth-order peptide approximations. The nth-order peptide approximations may in turn be used to predict secondary structure in unknown sequences.

Experimental

Two data bases have been used for this study. The first is the *Protein Sequence Data Tape 76* consisting of 767 protein sequences prepared by M. O. Dayhoff at the National Biomedical Research Foundation (32). The second data base is a subset of the AMSOM (Atlas of Macromolecular Structure on Microfiche) data base compiled at the National Institutes of Health by R. J. Feldmann (33).* The AMSOM subset contains fifty protein
*AMSOM is available from: Tracor Jifco, Inc., AMSOM program, 1776 E. Jefferson St., Rockville, MD 20852, USA.

sequences for which detailed secondary structure and protein sequence are known.

All computer programs were written in FORTRAN and all computations were carried out at the University of North Carolina Computation Center at Chapel Hill on the IBM Model 360 Series 75 and IBM Model 370 Series 155 computers.

Procedure

The basis of the information theoretical approach is that an nth-order peptide approximation can be made to represent the protein secondary structure. To test the validity of this premise frequencies of occurrence of various nth-order peptides were determined in the *Protein Sequence Data Tape 76*. The data base contains 76,094 classifiable peptides. There is a total of 77,267 residues in the data base; however, not all of these were useable in this study. Some of the peptides are designated as "unknown" or atypical (and are coded as UNK); others could not be distinguished between aspartic acid and asparagine (coded as ASX); and still others could not be distinguished between glutamic acid, glutamine and pyrrolidine-carboxylic acid (coded as GLX). These three residue types, UNK, ASX and GLX are considered unclassifiable in this study. As could be expected all possible twenty monopeptides are found to occur at least once in the data base; all 400 dipeptides also occur. (The order of a polypeptide is defined by starting at the amino end and proceeding towards the carboxy end.) There are 8,000 possible tripeptides and 89.60% are found at least one time in the entire data base. It is believed that this approaches a statistically valid result as there are only 8,000 possible tripeptide combinations in a data base of 72,363 classifiable tripeptides, (i.e., they do not contain UNK, ASX or GLX). Therefore it would appear that approximately 10% of the possible tripeptides are rarely occuring combinations in protein biochemistry. Tetrapeptides were also examined. From the 70,865 classifiable tetrapeptides in the data base, only 21.55% of the possible 160,000 tetrapeptides occur at least one time. This is not a statistically valid result as there are more than twice as many possible tetrapeptides as there are available in the data base. For this reason the present work was limited to the study of tripeptides. It is true that as the order of approximation is increased, the approximation more closely approaches the total information content of the message. However, if higher than third-order peptide (tripeptide) approximations are used, the approximations become seriously biased by the data base size.

Further investigation of the *Protein Sequence Data Tape 76* yielded that the most commonly occuring tripeptide is proline-proline-glycine. This residue occurs almost exclusively in collagen and keratin, both fibrous proteins. As the study involved only globular proteins the entire data base could not be adequately used and a globular subset of the *Protein Sequence Data Tape 76* was formed by removing all the fibrous proteins. This subset data base contains 65,783 classifiable peptides and 62,511 classifiable tripeptides. This is still a statistically valid data collection for the 8,000 possible tripeptide combinations. Statistics on the globular subset indicate that 87.85% of all possible tripeptides are present. Unfortunately, not all the conformations are known for all the protein sequences in the globular subset data base; therefore, this data base could only be used for determining the statistical make-up of tripeptides in globular proteins.

The AMSOM data base subset contains complete secondary structure and amino acid sequence for fifty proteins. There is a total of 9,702 classifiable peptides in the subset and 9,364 classifiable tripeptides. This data subset is not sufficiently large to yield a statistically valid result for tripeptide occurrence; however, from data on the most frequently occuring tripeptides it was ascertained that the AMSOM subset is an adequate representation of the *Protein Sequence Data Tape 76* globular subset. Once this had been demonstrated a third-order peptide approximation table of the AMSOM subset was constructed.

A single table to represent all the tripeptide frequencies would involve 8,000 x 27 (or 216,000) entries. There are 8,000 possible tripeptides - each of which may exhibit twenty-seven different conformational states. As tripeptides are selected from the protein sequences each individual peptide can exhibit either α, β or random coil (R) conformation. A list of all twenty-seven possible tripeptide conformations may be found in Table I.

Table I. Possible Tripeptide Conformations

RRR	RRα	RRβ	RαR	R$\alpha\alpha$	R$\alpha\beta$	RβR	R$\beta\alpha$	R$\beta\beta$
αRR	αRα	αRβ	$\alpha\alpha$R	$\alpha\alpha\alpha$	$\alpha\alpha\beta$	$\alpha\beta$R	$\alpha\beta\alpha$	$\alpha\beta\beta$
βRR	βRα	βRβ	$\beta\alpha$R	$\beta\alpha\alpha$	$\beta\alpha\beta$	$\beta\beta$R	$\beta\beta\alpha$	$\beta\beta\beta$

It was hoped that the size of the table could be effectively reduced as computer storage of such a large table is prohibitive even on a large machine. As an alpha helix cannot be defined for any fewer than four residues, the four tripeptide conformations that exclude any more than one alpha helical peptide are forbidden conformations. These four conformations are RαR, R$\alpha\beta$, $\beta\alpha$R and $\beta\alpha\beta$. Although a single beta sheet peptide is defined it was found that only the RβR tripeptide conformation occurs; whereas the other three, R$\beta\alpha$, $\alpha\beta$R and $\alpha\beta\alpha$, do not. All remaining twenty tripeptide conformations occur at least once in the data base. The table had thus been reduced to 8,000 x 20 (or 160,000) entries.

Rather than use the tripeptide frequency table in its present form it was decided to reduce it to 8,000 x 3 (or 24,000) entries. This was accomplished by determining the overall alpha helix, beta sheet and random coil tendencies of every tripeptide. For example, the various conformations of the tripeptide alanine-aspartic acid-alanine are shown in Figure 1(a). The tripeptide occurs eight times in the entire AMSOM subset: three times as RRR (or 37.5% of the time), twice as $\alpha\alpha\alpha$ (25.0%), twice as $\beta\beta\beta$ (25.0%), and once as βRR (12.5%). As this tripeptide has a single alpha conformation, $\alpha\alpha\alpha$, the total alpha contribution of the tripeptide is 0.250. As alanine-aspartic acid-alanine has two beta conformations, $\beta\beta\beta$ and βRR, its total beta contribution is 0.250 + 0.125 or 0.375. Similarly the total random coil contribution is 0.375 + 0.125 or 0.500. The final table entry for this tripeptide is shown in Figure 1(b). In this way the tripeptide conformations are biased towards the most likely conformation of the tripeptide.

An illustration of how this information is used to determine the structure of a single amino acid in a protein is shown in Figure 2. The pentapeptide glutamine-alanine-aspartic acid-alanine-asparagine was randomly extracted from a protein sequence. There are three tripeptides in the pentapeptide that contain aspartic acid and these appear in the first column of Figure 2. The method of prediction is to sum the alpha, beta and random coil contributions and use the maximum as an indicator for the conformation of the single peptide under investigation. As can be seen from Figure 2 the maximum is 1.700 for random coil conformation. If the tripeptides are individually examined one of the three clearly favors random coil, one favors alpha helix, and the third cannot distinguish between beta sheet and random coil. By this method aspartic acid would always be predicted random coil in this particular pentapeptide, but should any of the other four peptides be different, the prediction may also change.

Reduction of the tripeptide frequency table from twenty to three conformations must introduce some error. It has been found that approximately 78% of all the classifiable tripeptides in the AMSOM subset are either $\alpha\alpha\alpha$, $\beta\beta\beta$, or RRR. Thus for this 78% of all conformations the reduction to alpha, beta or random coil contributions makes no difference in this assignment. An additional 21% of the tripeptides have two of the peptides the same and one different. (A complete list is given in Figure 3.) Therefore, the assignment of overall conformation within this 21% to a single peptide in a given tripeptide is correct for an additional 14% of all conformations (2/3 x 21% = 14%). Consequently the overall conformation approximation is correct for 92% (78% + 14%) of all conformations in the data base subset.

Results and Discussion

The above prediction scheme was performed on all fifty

(a)   ADA

RRR	ααα	βββ	βRR
0.375	0.250	0.250	0.125

(b)   ADA

α	β	R
0.250	0.375	0.500

Journal of Biological Chemistry

Figure 1.   Reduction of the table entry of alanine–aspartic acid–alanine from 20 potential conformations to three. Four of the potential 20 entries occur (a) and are reduced to three by summing all conformations that contain α-helices, all that contain β-sheets, and all that contain random coils to give total α-tendency, total β-tendency, and total random coil tendency (b). The codes A, D, A are the IUPAC–IUB single letter codes for the amino acids (34).

... QADAN ...

	α	β	R
QAD	0.0	1.000	1.000
ADA	0.250	0.375	0.500
DAN	0.800	0.0	0.200
	1.050	1.375	1.700

Figure 2.   The conformation for the amino acid, aspartic acid, is determined in the pentapeptide glutamine–alanine–aspartic acid–alanine–asparagine by summing the α, β, and random coil tendencies of the three tripeptides that incorporate aspartic acid. In this pentapeptide aspartic acid is assigned a random coil configuration as the sum of the random coil tendencies is the maximum.

proteins that comprised the AMSOM subset. Two corrective procedures were added to the prediction algorithm. First, all single
peptide alpha helices were disallowed and converted to either
random coil, or beta sheet only if preceded and succeeded by beta
sheets. Second, all single peptide beta sheets were disallowed
except those preceded and succeeded by random coil peptides. In
order to measure how well the prediction algorithm correlated
with the observed structure, two parameters were computed for each
prediction. The first of these two parameters, Recall, is defined as follows:

$$\text{Recall} = \frac{\text{Total Number of Conformation Predicted Correctly}}{\text{Total Number of Conformation Observed}}$$

That is to say, the numerator is the number of peptides found to
have the correct conformation after prediction. The denominator
is the total number of peptides of the specific conformation
observed in the protein and known to be correct. Recall does not
account for over-predictions. For example, if a protein is known
to contain only alpha helices and random coil prior to prediction,
the Recall can be forced to 100% by predicting the entire sequence
alpha helical. The second parameter, Precision, is a measure of
over-prediction, and is defined as:

$$\text{Precision} = \frac{\text{Total Number of Conformation Predicted Correctly}}{\text{Total Number Predicted to be in Conformation}}$$

The numerator is the same as in the Recall, but the denominator
is equivalent to the total number that were predicted by the
algorithm to be in the specific conformation. Hence, a Precision
of 100% indicates no over-prediction, and as the over-prediction
increases the Precision decreases. The results of the correlation between prediction and observation of alpha helices and beta
sheets for all fifty proteins in the AMSOM subset are listed in
Table II.

Table II shows that the predictive algorithm is quite successful. The Recall for alpha and beta conformations is generally greater than 75% and the Precision is generally greater than
80%. The overall results for the entire data base show that the
Recall and Precision for alpha helices are 82.6% and 93.7%,
respectively; whereas, the same values for beta sheets are 83.6%
and 90.6%, respectively. One recent study (30), shown to be at
least as accurate as those preceding it, yielded an average Recall
of 68.6% for alpha helices and an average Recall of 53.1% for beta
sheets. Precision, per se, was not calculated. The information
theoretical approach yielded average Recalls of 79.3% and 83.7%
for alpha helices and beta sheets, respectively. These results
would indicate that the information theoretical approach is more
accurate than other techniques; however, these correlations were
determined using a small data set of only 9,364 classifiable

Table II. Recall and Precision

Protein Name	α		β	
	Re-call%	Preci-sion %	Re-call%	Preci-sion %
Lamprey Cyanmethemoglobin	85.5	100.0	n/a*	n/a
Horse Aquomethemoglobin Dimer	82.5	96.9	n/a	n/a
Human Deoxyhemoglobin Dimer	92.1	97.7	n/a	n/a
Marine Worm Carboxyhemoglobin	88.8	96.3	n/a	n/a
Bovine Ferricytochrome b5	80.8	100.0	95.8	82.1
Tuna Ferrocytochrome c	88.5	97.9	n/a	n/a
Tuna Ferricytochrome c "outer"	88.5	97.9	n/a	n/a
Bacterial Ferricytochrome c2	89.4	95.5	n/a	n/a
Bonito Ferrocytochrome c	89.1	87.2	n/a	n/a
Sperm Whale Metmyoglobin	83.5	98.1	n/a	n/a
Bacterial Rubredoxin	n/a	n/a	n/a	n/a
Bacterial High Potential Iron Protein	55.6	71.4	82.4	100.0
Bacterial Ferredoxin	n/a	n/a	n/a	n/a
Bacterial Thioredoxin	85.7	92.3	81.3	86.7
Marine Worm Hemerythrin	77.8	95.5	n/a	n/a
Fungal Ferrichrysin	n/a	n/a	n/a	n/a
Subtilisin BPN'	75.6	100.0	83.8	72.1
Bovine Tosyl Alpha-Chymotrypsin A	87.5	100.0	93.9	95.9
Bovine Chymotrypsinogen A	53.9	87.5	93.0	95.9
Bovine Gamma-Chymotrypsin A	100.0	78.8	93.0	95.9
Bovine Trypsin	72.7	80.0	87.2	90.1
Porcine Tosyl Elastase	100.0	84.6	86.1	89.2
Bacterial Serine Protease b	75.0	81.8	79.2	83.6
Papain	83.3	92.6	71.9	71.9
Bacterial Thermolysin	79.0	98.9	84.9	80.5
Bovine Carboxypeptidase a Complex	86.1	99.0	84.4	74.1
Bovine Carboxypeptidase b	84.0	96.7	88.9	78.4
Bovine Trypsin Inhibitor	72.3	88.9	100.0	91.3
Dogfish Apo-lactate Dehydrogenase	76.2	95.2	88.6	87.5
Lobster Glyceraldehyde-3-P Dehydrogenase "Red"	87.1	88.0	82.7	96.5
Horse Alcohol Dehydrogenase Complex	80.2	93.9	77.7	95.9
Bacterial Oxidized Flavodoxin	82.5	90.4	81.0	94.4
Bovine Ribonuclease S Complex	93.8	100.0	93.0	90.9
Bacterial Nuclease Complex	88.9	78.1	81.3	95.1
Mouse Immunoglobin A Fab "MCPC#603"	n/a	n/a	80.2	97.1
Human Bence-Jones Protein Dimer "REI"	n/a	n/a	86.3	90.0
Human Immunoglobin G Fab' "New"	n/a	n/a	86.0	93.5
Human Bence-Jones Dimer "MCG"	n/a	n/a	87.1	97.1
Porcine Adenylate Kinase	91.5	97.9	95.8	85.2
Yeast Phosphoglycerate Kinase	29.0	71.4	47.3	74.3
Jack Bean Concanavalin A	n/a	n/a	80.5	95.7
Chicken Lysozyme	67.6	100.0	58.3	63.6

Table II. Recall and Precision (cont.)

Protein Name	α		β	
	Re-call%	Preci-sion %	Re call%	Preci-sion %
Chicken Triose Phosphate Isomerase (Monomer 1)	86.3	95.2	76.0	92.7
Carp Calcium Binding Protein B	82.3	98.1	100.0	81.8
Bovine Cu, Zn Superoxide Dismutase	n/a	n/a	74.1	92.3
Human Carbonic Anhydrase B	78.7	97.4	85.1	93.8
Human Carbonic Anhydrase C	83.0	95.1	79.1	93.6
Human Prealbumin Dimer	87.5	53.9	81.0	97.1
Sea Snake Neurotoxin	n/a	n/a	93.8	75.0
Sea Snake Neurotoxin	n/a	n/a	64.5	90.9
Overall Results	82.6	93.7	83.6	90.6

*n/a = not applicable, i.e., protein did not exhibit conformation.

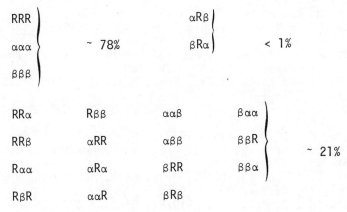

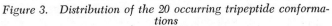

Figure 3. Distribution of the 20 occurring tripeptide conformations

tripeptides. It is not known whether this data base adequately represents all protein sequences; hence, prediction of alpha helices and beta sheets using these data may not yield reliable results for unknown protein structures. This uncertainty may be resolved with a larger, more representative data base.

Acknowledgement

The authors would like to thank R. J. Feldmann of the National Institutes of Health for supplying the AMSOM data base subset. The support of the National Science Foundation is gratefully acknowledged.

Literature Cited

1. Guzzo, A. V., Biophys. J. (1965), 5, 809-822.
2. Prothero, J. W., Biophys. J. (1966), 6, 367-370.
3. Periti, P. F., Quagliarotti, G., and Liquori, A. M., J. Mol. Biol. (1967), 24, 313-322.
4. Cook, D. A., J. Mol. Biol (1967), 29, 167-171.
5. Schiffer, M. and Edmundson, A. B., Biophys J. (1968), 8, 29-39.
6. Dunnill, P., Biophys J. (1968), 8, 865-875.
7. Low, B. W., Lovell, F. M. and Rudko, A. D., Proc. Natl. Acad. Sci. U.S. (1968), 60, 1519-1526.
8. Kotelchuck, D. and Scheraga, H. A., Proc. Natl. Acad. Sci. U.S. (1969), 62, 14-21.
9. Lewis, P. N., Gō, N., Gō, M., Kotelchuck, D. and Scheraga, H. A., Proc. Natl. Acad. Sci. U.S. (1970), 65, 810-815.
10. Ptitsyn, O. B. and Finkel'shtein, A. V., Dohl. Akad. Nauk. S.S.S.R. (1970), 195, 221-224.
11. Ptitsyn, O. B. and Finkel'shtein, A. V., Biofizika (U.S.S.R.), (1970), 15, 757-767.
12. Leberman, R., J. Mol. Biol. (1971), 55, 23-30.
13. Robson, B. and Pain, R. H., J. Mol. Biol. (1971), 58, 237-259.
14. Lewis, P. N., Momany, F. A., and Scheraga, H. A., Proc. Natl. Acad. Sci. U.S. (1971), 68, 2293-2297.
15. Wu, T. T. and Kabat, E. A., Proc. Natl. Acad. Sci. U.S. (1971), 68, 1501-1506.
16. Finkel'shtein, A. V. and Ptitsyn, O. B., J. Mol. Biol. (1971), 62, 613-624.
17. Nagano, K., J. Mol. Biol (1973), 75, 401-421.
18. Kabat, E. A. and Wu, T. T., Proc. Natl. Acad. Sci. U.S. (1973), 70, 1473-1477.
19. Kabat, E. A. and Wu, T. T., Proc. Natl. Acad. Sci. U.S. (1974), 71, 4217-4220.
20. Nagano, K., J. Mol. Biol. (1974), 84, 337-372.
21. Burgess, A. W., Ponnuswamy, P. K. and Scheraga, H. A., Isr. J. Chem. (1974), 12, 239-286.

22. Chou, P. Y. and Fasman, G. D., Biochemistry (1974), 13, 211-222.
23. Chou, P. Y. and Fasman, G. D., Biochemistry (1974), 13, 222-245.
24. Lim, V. I., Biofizika (1974), 19, 366-378.
25. Lim, V. I., Biofizika (1974), 19, 562-575.
26. Lim, V. I., J. Mol. Biol. (1974), 88, 857-872.
27. Lim, V. I., J. Mol. Biol. (1974), 88, 873-894.
28. Schulz, G. E., Barry, C. D., Friedman, J., Chou, P. Y., Fasman, G. D., Finkel'shtein, A. V., Lim, V. I., Ptitsyn, O. B., Kabat, E. A., Wu, T. T., Levitt, M., Robson, B. and Nagano, K., Nature (1974), 250, 140-142.
29. Matthews, B. W., Biochim. Biophys. Acta. (1975), 405, 442-451.
30. Argos, Patrick, Schwarz, James and Schwarz, John, Biochim. Biophys. Acta. (1976), 439, 261-273.
31. Shannon, C. E. and Weaver, W., "The Mathematical Theory of Communication," pp. 43-44, University of Illinois Press, Urbana, Illinois, 1972.
32. Dayhoff, M. O., "Atlas of Protein Sequence and Structure," vol. 5, National Biomedical Research Foundation, Washington D.C., 1972; vol. 5., suppl I, 1973; vol. 5., supp. 2, 1976.
33. Data base supplied on magnetic tape from Richard J. Feldmann, personal communication.
34. IUPAC-IUB Commission on Biochemical Nomenclature, J. Biol. Chem. (1968), 243, 3557-3559.

5

Computer-Assisted Structure Elucidation Using Automatically Acquired ^{13}C NMR Rules

GRETCHEN M. SCHWENZER and TOM M. MITCHELL

Department of Computer Science, Stanford University, Stanford, CA 94305

Carbon-13 nuclear magnetic resonance (CMR) has developed into an important tool for the structural chemist. A CMR spectrum exhibits a wide range of shifts which have been shown to have a strong correlation with structure(1,2). A natural abundance CMR spectrum which is fully proton decoupled consists of a number of sharp peaks which correspond to the resonance frequencies in an applied magnetic field of the various types of carbon atoms present. A C-13 shift is the amount an observed peak is shifted from that of a reference peak, usually tetramethylsilane (TMS).

Molecular structure elucidation using CMR consists of establishing a set of rules which summarize the CMR behavior for a set of compounds and then using the rules to identify unknown compounds. In the traditional approach to structure elucidation using CMR the chemist forms a set of empirical rules by sorting through a large amount of data looking for correlations between structural arrangements in the molecules and the observed C-13 shift. The total shift is then given as a function of these structural parameters. The functional form is usually chosen to be a linear combination of independent parameters. The optimized value of the coefficient of each structural parameter is obtained by a curve fitting procedure. This approach leaves the decisions of the structural parameter selection and the selection of a reasonable functional form to the chemist. In both cases the correct decisions are easily overlooked. The best known example of this approach is that of Lindeman and Adams(3).

Although the rule form resulting from the parameter approach is useful for predicting spectra of a given structure it cannot be used efficiently for

58

structure elucidation. Observing a shift does not give any information about what substructure is present without constructing a set of candidate structures, calculating their shifts and then comparing the calculated shifts with the observed spectrum. This procedure was followed in the work of Carhart and Djerassi (4) which used the parameter set obtained by Eggert and Djerassi for the acyclic amines (5). However, this method does not reflect the actual thinking procedure a chemist would use when identifying an unknown and it suffers from its inability to be generalized. The selection of empirical rules and the algorithms for the generation of substructures are specific for each class of compounds. To the chemist it is more likely that the appearance of an observed shift or set of shifts would suggest a structural arrangement.

A second approach to structure elucidation is to establish a data base of C-13 NMR spectra and then use the data base to do structure elucidation by searching for peak patterns similar to those in the unknown spectrum(6,7,8). A data base which consists of the entire spectrum results in storage of a large amount of redundant information. Structure elucidation is accomplished by supplying the chemist with the molecules having portions of their spectra which are similar to that of the unknown. The chemist then uses these to suggest partial substructures that are present in the unknown. This method has suffered from the necessity of obtaining large data bases and methods of assembling the partial substructures.

We offer an alternative to these approaches with the following procedure. First rule formation is accomplished by a computer program which forms a set of empirical rules by associating an observed total shift with a substructural arrangement in the molecule. A second program uses these rules for structure elucidation by assembling substructural fragments suggested by the rules.

A sample rule constructed by the program from a set of paraffins and acyclic amines is

$$CH_3^* -CH_2-CH_2-CH_2- \quad ---> \quad 14.0ppm \leqslant \delta(*) \leqslant 14.7ppm.$$

The asterisk denotes the atom for which the shift is predicted. The prediction $\delta(*)$ is given in ppm downfield from TMS. The important substructural arrangements given in the rule are actually constructed by the program from a language of features supplied by the user. Molecular structure elucidation is accomplished

by observing a total shift and finding the rules which
are possible explanations for the shift.   The rules
selected postulate partial substructures which might be
in the molecule. These substructures are assembled to
construct the final molecule. A description of the
rule formation and structure elucidation programs
applied to the paraffins and acyclic amines is given in
the following sections. We believe the algorithms used
are general enough to treat widely different classes of
compounds.   Rules   generated   for   decalins,
methyldecalins and hydroxy-steroids are shown in the
third section.

Empirical Rule Formation

    Rule Generation.   The rule generation program(9)
must be supplied a training set of known structures
with their assigned spectra. A set of primitive terms
which will form the language of atom features used to
describe the atoms and bonds (atom type, number of non-
hydrogen   neighbors,   orientation   of   substituents,
etc.)must also be supplied. These terms are combined
to construct structural fragments which imply a total
shift.   The chemist also sets two parameters which
regulate the   generality of the   rules generated.
MINIMUM-EXAMPLES is a parameter which specifies the
minimum number of data points which a rule must explain
within the training set. The other parameter, MAXIMUM-
RANGE, specifies the maximum allowable shift range for
a rule.   If the chemist wants only the most general
trends in the data he can require a larger number of
examples with moderately sized shift ranges.
    The format of the rules generated is

$$\text{substructure} \xrightarrow{\text{implies}} {}^{13}C \text{ shift range.}$$

If the substructure to the left of the arrow is present
within some molecule then there is a shift within the
range given to the right of the arrow. The rule shown
in Table I was generated on a combined set of acyclic
amines and paraffins.

Table I   Rule Form

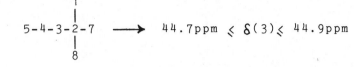

$$5-4-3-2-7 \longrightarrow 44.7\text{ppm} \leqslant \delta(3) \leqslant 44.9\text{ppm}$$

Node	Atom Type	Number of non-hydrogen Neighbors
1,5,7,8	C	$\geqslant 1$
2	C	4
3	C	2
4	C	2

For the substructure shown in Table I with the corresponding atom features for atom type and number of non-hydrogen neighbors atom number 3 will have a C-13 shift in the range 44.7ppm to 44.9ppm downfield from TMS.

The rule search procedure is shown in Figure 1. The search begins with the general seed rule $C->-\infty \leqslant \delta_c \leqslant \infty$ (where C may be any carbon atom with any atom properties and is the observed shift) and proceeds to expand this rule by adding new atoms and atom features to the substructure which will narrow the predicted range of shifts. The seed rule in Figure 1 is expanded by considering all possible values of "number of neighbors" of the central carbon. Each resulting level 1 substructure is expanded in level 2 by adding either an "atom type" or "number of neighbors" specification to each atom one bond away from the central carbon. At each step only a single atom feature from the user supplied list is added. Each substructure generated is associated with a range of C-13 shifts. This range is determined by searching for occurrences of the substructure within the training set molecules. The shift range associated with the substructure is the range of all occurrences of the substructure in the training set.

Each substructure generated in the rule search is evaluated in terms of the associated shift range. If the shift range is narrower than the range of the parent rule then the added specification is considered to be useful and the search continues from the new substructure, otherwise the path is terminated. The

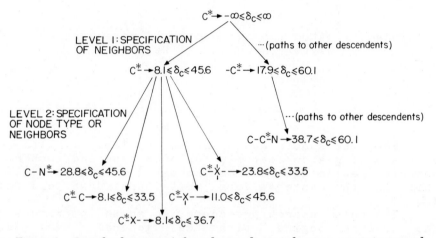

Figure 1. Partial schematic of the rule search. δ_c *values are approximate and are given in ppm downfield from TMS. (*) identifies the carbon atom to which the shift is assigned; (X) indicates any non-hydrogen atom.*

program runs until all branches of the search have been explored. For a substructure to be accepted as a final rule it must satisfy the conditions of MINIMUM-EXAMPLES and MAXIMUM-RANGE.

Since the rule search procedure can result in slightly different substructures being generated which cover the same data in the training set, the generated rule set may be redundant. A less redundant set of rules is chosen by assigning a score to each rule. The score is peaks/range^2 where peaks is the number of data peaks covered by the rule in the training data, and range is the width of the shift range for the rule. The rule with the highest score is selected and all the data points explained by that rule are removed. If unexplained data points remain, the score of the remaining rules are reevaluated and the procedure repeated. The intent is to select the strongest rule during each iteration and weaken the rules with evidence which overlap with it. The result is a subset of the strongest rules covering the same data as the original rule set

The algorithm used to generate the C-13 NMR rules is similar to the algorithm in the Meta-DENDRAL program which generates empirical rules of molecular fragmentation from mass spectral data(10,11).

For a set of 22 paraffins and 47 acyclic amines with a total of 435 data peaks a set of 138 rules were generated with the parameter settings MINIMUM-EXAMPLES=2 and MAXIMUM-RANGE=2.0 ppm.

Structure Selection. To test the information content of the rules and thus their value for use in structure elucidation, a structure selection test was designed. A program was written which uses the rules to predict spectra for a set of candidate molecules and compares the predicted spectra to that of an unknown. The candidates are then ranked according to a "best" match criteria to the unknown spectrum.

A spectrum is predicted by applying the rules to a molecule, searching for places the rule substructure fits into the molecule. This is done with a graph matching routine. When a match is found the shift range associated with the rule is predicted for the associated carbon atom. Figure 2 is a example of predicting a spectrum. Each rule shown has a substructure with corresponding atom features that maps into the molecule. Often more than one rule applies to a given atom to give different predictions. If the predicted ranges are consistent(i.e. one of the predicted ranges is contained in the others) the

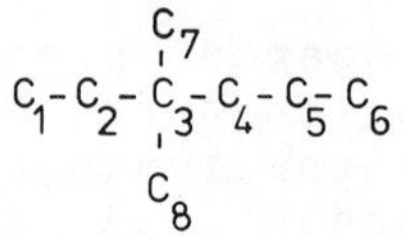

$$C_1-C_2-\overset{\overset{\displaystyle C_7}{|}}{\underset{\underset{\displaystyle C_8}{|}}{C_3}}-C_4-C_5-C_6$$

OBSERVED SPECTRUM

| 8.7 | 15.4 | 17.9 | 27.1 | 33.4 | 34.9 | 44.9 |

PREDICTED SPECTRUM

(8.1 8.7) (14.0 15.4) (17.9 24.3) (27.1 29.7) (29.7 35.6) (31.2 34.9) (44.7 44.9)

CARBON ATOM   RULES WHICH APPLY TO CARBON ATOM

	SUBSTRUCTURE	ATOM NODE	ATOM TYPE	NUMBER OF NON-H NEIGHBORS	PREDICTION
C_4	$\begin{array}{c}1\\|\\5\text{-}4\text{-}3\text{-}2\text{-}7\\|\\8\end{array}$	1,5,7,8	C	≥1	44.7 ≤ δ(3) ≤ 44.9
		2	C	4	
		3	C	2	
		4	C	2	
	$-4\text{-}3\overset{|}{\underset{|}{\text{-}2}}-$	2	C	4	41.1 ≤ δ(3) ≤ 45.1
		3	C	2	
		4	C	2	
	1-2-3	1,3	C	≥1	17.9 ≤ δ(2) ≤ 56.9
		2	C	2	
C_3	$\begin{array}{c}3\\|\\8\text{-}2\text{-}1\\|\\7\end{array}$	1,3,7,8	C	≥1	29.7 ≤ δ(2) ≤ 35.6
		2	C	4	

Figure 2. Partial spectrum predictions for 3,3-dimethylhexane. δ(n) is the shift for atom n in ppm downfield from TMS.

narrowest predicted range is used. This is illustrated
by the rules which apply to C4 in Figure 2. The
predicted range 44.7 to 44.9 is contained in the other
two rule predictions thus it is selected as the
prediction. This method can be rationalized since the
actual shift should fall into all of the predicted
ranges and thus into the narrowest. If the predicted
ranges overlap incompletely or are disjoint then the
ranges are merged to arrive at a final predicted range
for the carbon atom.

   The predicted spectrum is compared to an unknown
spectrum by assigning each atom´s predicted range to
the closest observed shift in the unknown spectrum. In
order to be a valid assignment the number of carbons in
the structure less the number of observed shifts must
be greater than or equal to the number of multiply
assigned shifts. If the assignment does not satisfy
this criterion the required number of multiply assigned
observed shifts are reassigned. The atoms which are
chosen to be reassigned are those whose reassignment
will cause the smallest change in the comparison score
assigned to the match of the predicted spectrum to the
unknown spectrum.

   Results. A set of rules has been generated using
a subset of the paraffin data from Lindeman and Adams
(3) and a subset of the acyclic amine data from Eggert
and Djerassi (5). Molecules with the empirical formula
C_9H_{20} and $C_6H_{15}N$ were excluded from the training set.
The structure selection test was performed by
generating all structural isomers with the empirical
formulas C_9H_{20} (35 isomers) and $C_6H_{15}N$ (39 isomers).
For each of the candidate isomers a spectrum was
predicted. There were 24 C_9H_{20} spectra available from
the work of Lindeman and Adams. The 35 predicted
spectra were compared and ranked against each of these
available spectra. The results of this ranking for
C_9H_{20} as well as a similar test on $C_6H_{15}N$ are shown in
Table II.

Table II.   Results of Structure Selection

Empirical Formula	Number of Candidates	Number of Structures Ranking			
		1st	2nd	$...6^{th}$	$..9^{th}$
C_9H_{20}	35	$\frac{20}{24}$	$\frac{3}{24}$		$\frac{1}{24}$
$C_6H_{15}N$	39	$\frac{8}{11}$	$\frac{2}{11}$	$\frac{1}{11}$	

Peak intensity information which gives the number of carbon atoms corresponding to an observed peak was not used. The use of this information would have resulted in the correct assignment for those which were poorly ranked.

The form of the rule which is proposed has several advantages. Each rule has its own predicted error and the rule set consists of rules of varying detail. In the parameter set approach to rule formation the error is the standard deviation of the training set data from the fitted curve. When analyzing carbon atoms that exhibit magnetic nonequivalence (resulting in different chemical shifts for two identical groups in molecules having an asymmetric carbon atom) the parameter set approach which attempts to predict the arithmetic mean will always be in error. The advantage in predicting total shifts over hypothesizing partial contributions is in avoiding initial biases as to what contributes to the shift and how these contributions are to be combined. Our program bypasses the bias of assumed functional form and introduces only a weak bias concerning which structural features may be considered. The program can consider any structure which can be constructed from the atom feature language supplied by the chemist. Another advantage of the rule format is that it can be read backwards, that is, the appearance of a peak in an unknown spectrum implies a structural feature. This property is what distinguishes the efficiency of this rule form over other forms when used for structure elucidation.

The method of structure selection described above

is inefficient when there are a large number of structural isomers. Instead of applying this test to all structural isomers for a given empirical formula we wish to select a subset of likely candidates to be put through this final ranking procedure. The ability to read the rules backwards to do molecular structure elucidation will enable us to achieve this goal.

Structure Elucidation

   Structure Search.   The information the chemist must supply to the structure elucidation program includes the empirical formula of the unknown and the observed spectrum. Two parameters must also be set by the chemist. The first parameter is the number of plausible structures which should be found to be ranked by the structure selection procedure described earlier. The second parameter is the error range in ppm which should be assigned to the rules to account for deficiencies in the training set, experimental error, solvent effects, etc.
   In order to make the generated rules more useful for structure elucidation additional information is added to the rules. The form of the rule used for structure elucidation is shown in Table III.   The rule now says the observation of a shift implies a partial substructure.   This is the same rule as shown in Table I with the addition of two new properties, support prediction and secondary prediction.

Table III Rule Form for Structure Elucidation

$$44.7\text{ppm} \leqslant \delta(3) \leqslant 44.9\text{ppm} \ \text{----}> \ 5\text{-}4\text{-}3\text{-}2\text{-}7$$

with branches: node 1 attached above node 2, node 8 attached below node 2.

Node	Atom Type	Number of non-H Neighbors	Support Prediction	Secondary Prediction
1	C	$\geqslant 1$		$27.1 \leqslant \delta(1) \leqslant 34.9$
2	C	4	$29.7 \leqslant \delta(2) \leqslant 35.6$	$30.7 \leqslant \delta(2) \leqslant 33.4$
3	C	2		
4	C	2	$17.9 \leqslant \delta(4) \leqslant 56.9$	$17.9 \leqslant \delta(4) \leqslant 27.6$
5	C	$\geqslant 1$		$15.4 \leqslant \delta(5) \leqslant 24.3$
7	C	$\geqslant 1$		$27.1 \leqslant \delta(7) \leqslant 34.9$
8	C	$\geqslant 1$		$27.1 \leqslant \delta(8) \leqslant 34.9$

Once a set of rules has been obtained from the rule generation program, support predictions can be made. For each rule an attempt is made to find support predictions for all atoms in the rule substructure except the atom for which the rule is defined. For each rule's substructure the rules are applied to it to obtain predictions. Predictions made by the rule set for any atom in the substructure are tabulated. The final support prediction for a particular atom is obtained by merging the tabulated predictions for that atom. Support predictions are really main predictions merged into the rule. This is done to introduce additional constraints upon the selection of the rules early in the search procedure. In the rule shown in Table III there were other rules which gave predictions for nodes 2 and 4.

Although support predictions cannot be obtained for all atoms in the rule substructure, predicted ranges are associated with all atoms using the following procedure. Secondary predictions are made by finding all places the rule substructure applies in the training set data and tabulating the observed shifts for the atoms in the substructure. For each atom in the substructure the observed shifts found in the training set are merged to form the secondary predictions. The reliability of a secondary prediction is highly dependent on the variety of molecular structures in the training set. Secondary predictions for atoms which are fewer bonds away from the rule's predicted atom will have a smaller expected error than those more bonds away. The shift ranges of the secondary predictions are broadened by a predefined constant to account for the deficiencies in the training set. The addition of support and secondary predictions to the original rules will form the new rule set.

The first step in the rule search is to select a subset of the rule set which will act as possible explanations for the observed spectrum. The rules which have shift ranges consistent with the shifts in the observed spectrum are selected. Each rule selected is checked for agreement with the empirical formula of the unknown molecule and for the presence of observed peaks in the spectrum for all support predictions in the rule. The constraint that the number of carbons in the structure less the number of observed shifts must be greater than or equal to the number of multiply assigned shifts must also be satisfied. For the rule to be a valid possible explanation there must be an assignment of the main and support predictions which does not violate this constraint. The subset of rules selected by this procedure from the set of partial substructure hypotheses for the observed spectrum.

The structure search procedure is shown in terms of a specific example in Figure 3. The observed spectrum is for a molecule with the empirical formula C_8H_{18}. The number of rules which are possible explanations for the observed shifts are shown. The observed spectrum in Figure 3 corresponding to 3,3-dimethylhexane was also shown in Figure 2. In Figure 2 only the rules which were correct explanations for the molecule are shown. For example there was one correct explanation for C3 explaining the observed shift 33.4. In Figure 3 there are eight possible explanations for 33.4 of which only one is correct for this particular spectrum.

An approximate lower bound to the number of possible ways the partial substructures may be assembled is the product of the number of explanations given for the observed shifts. Although in this example this lower bound is greater than 1000, the constraints imposed by the observed spectrum, empirical formula and rule set directed the search to the correct structure after considering only 12 paths through the search tree.

The search strategy shown in Figure 3 is that of a depth first tree search. Solutions exist at unknown locations in the tree. A set of heuristics or judgmental rules are chosen to guide the search to the final structure in the most efficient manner.

The subset of rules which are possible explanations for the observed spectrum are ordered to select the rule which has the greatest chance of being in the molecule and which will lead most rapidly to the final molecule. The heuristics which order the rules include the quality of the main and support predictions. The quality of a prediction means the width of the prediction range and the closenesss of the observed peak to the rule prediction range. The number of peaks explained in the training set by the rule is also a factor in ordering the rules. An analogy can be drawn between this heuristic and the selection by the chemist of a substructure which has frequently been observed to be present when a particular shift occurs in a spectrum. Another heuristic is the number of other explanations a particular rule has. If there is only one explanation for an observed shift it is very likely that the rule's substructure will be in the unknown molecule. These conditions could be considered

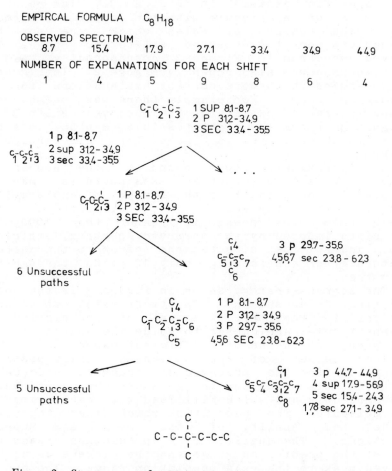

EMPIRCAL FORMULA $C_8 H_{18}$

OBSERVED SPECTRUM
  8.7   15.4   17.9   27.1   33.4   34.9   44.9

NUMBER OF EXPLANATIONS FOR EACH SHIFT
    1   4   5   9   8   6   4

Figure 3. Structure search. Main (P), support (SUP), and secondary (SEC) predictions are given in ppm downfield from TMS. The candidate structures are shown in capitals; the rules selected to build on the candidate structure are shown in lower case.

as contributions to the certainty that the rule substructure is in the final molecule.

The final heuristic which orders the subset of rule explanations will depend on the width of the secondary and support predictions. Instead of being a certainty measure of the rule substructure's presence it will be a measure of efficiency of building with this particular substructure. The neighbors of the rule's predicted atom are examined as potential building sites. The subset of neighbors which have more than one non-hydrogen neighbor will be the atoms on which construction will take place. The support and secondary predictions of these atoms suggest a subset of rules from those selected as explanations for the observed spectrum. This subset will form explanations for the shifts of these atoms. The narrowness of the support or secondary prediction determines the number of rules which are possible explanations. The rule which has the narrowest prediction for a neighboring atom will lead most rapidly to the final structure or the elimination of the rule as a possible explanation.

The combination of these heuristics will order the set of rule explanations, hopefully placing those which are the correct explanations with the most promising structures to build on early in the list. The rule chosen which satisfies these conditions for the example is shown in Figure 3. The substructure C1-C2-C3- has a support prediction for C1, a main prediction for C2 and a secondary prediction for C3.

Once a rule has been selected it is necessary to choose an atom in the rule to build on. Neighbors of atoms with main predictions and which do not have main predictions themselves will be the atoms which are possible candidates for building. First the support predictions for these atoms are considered. If any of the support predictions are sufficiently narrow that atom will be chosen as the construction site otherwise the secondary predictions are examined. The atom with the narrowest secondary prediction range is chosen. For the example in Figure 3 atom C1 was chosen for the site of construction. Rules are selected from the subset of rule explanations which have shift ranges consistent with the secondary or support prediction of the atom chosen. The set of rules selected is ordered with the most likely rule first using the criterion stated previously. The example in Figure 3 chose C1 to build on and the rule explanation for C1 is shown.

Two rules have now been chosen. The process of selection of the second rule results in an intial mapping between the two rule substructures. The atom from the first rule which was selected as the construction site maps to the atom with the main prediction in the second rule. It is now necessary to find all substructures consistent with the initial mapping that can result from the overlap of the rules' substructures. A graph matching routine is used to find the overlaped substructures. The properties checked in overlapping are those which make up the language of atom features. For the paraffins and acyclic amines properties checked are atom type and number of non-hydrogen neighbors. If no possible overlap exists a new rule explanation is tried, if there are no possible explanations for the selected atom the original rule chosen is discarded and the next choice is tried. Any candidate structures which result from the overlap must be consistent with the empirical formula of the unknown.

The overlap process must also merge the predictions for the two rules. An atom from one rule which has been mapped into an atom in the other rule must have a resultant prediction consistent with both predictions from the original rules. For instance the merging of two support predictions results in a prediction range which is the intersection of the two predictions. For the candidate substructure to be valid there must be an observed shift which is consistent with the new main and support predictions. In addition there must be an assignment of observed shifts to main and support predictions which satisfies the constraint of the number of multiply assigned shifts. In Figure 3 the overlapping of the first two rules does not result in any change in the substructure but the predictions of the two rules are merged.

The list of candidate structures which survive these tests have the same form as the original rules. The decision as to the likelihood of these candidate structures being in the molecule can be made on the basis of the candidate structure itself without combining the likelihood of the parent rules. The candidate structures are ranked on the basis of the quality of the main and support predictions.

The new subproblem is now identical to the original problem and the steps of selecting an atom to build on, selecting rule explanations for that atom, overlapping the two substructures, merging predictions, and the ranking of the resultant candidates are repeated. The procedure terminates when the required number of plausible molecules are found. Figure 3 shows the paths examined until the first plausible structure was found which was the correct solution for the observed spectrum. An unsuccessful path in Figure 3 means that the overlapping of a rule with the candidate structure failed on structural grounds or the candidate structures generated were not consistent with the observed spectrum. If a rule fails another rule is tried until a candidate structure is generated which is consistent with the constraints of the observed spectrum and empirical formula of the unknown molecule.

Results. One possible search procedure has been explained whose goal is to obtain a few plausible structures which can then be ranked by the structure selection procedure described earlier. Other possible search strategies are being considered whose goal would be to obtain substantial sized substructures which would then be used as starting points in a structure generating program such as CONGEN(12).

Handling Stereochemistry

The work on the paraffins and acyclic amines requires only topological descriptors in the language of atom features. Because of the dependence of C-13 shifts on stereochemical features (13,14) it is necessary to have the facility to include stereochemical terms when they are required. Substituents placed on systems which have static conformations such as trans decalin and androstane with trans ring fusions can be described in discrete terms. The terms we selected describe the orientation on the ring of the substituent as either axial or equatorial, and either alpha or beta. A substituent is beta in 10-methyl-trans-decalin if it is on the same side of the ring as the methyl group and alpha if on the opposite side of the ring from the methyl group. The rule generation program with the extension of the language to include these atom features was run on a combined set of trans decalins, 10-methyl-trans-decalols and monohydroxylated androstanes with trans ring fusions selected from the works of Grover and Stothers (13) and Eggert et. al.(14). Sixty rules were generated to cover the 249 data peaks of 17 compounds. Samples of the rules generated are shown in Figure 4. The examination of these rules will show that they are useful for the chemist who wants to study contributions to the total shift as well as for the structure elucidation procedure we have outlined.

In Figure 4 are shown two pairs of rules for alpha carbons. Within each pair of alpha carbon rules there is little shift difference between the axial or

Alpha Carbon Rules

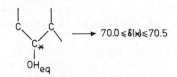 → $70.0 \leqslant \delta(*) \leqslant 70.5$

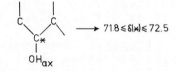 → $71.8 \leqslant \delta(*) \leqslant 72.5$

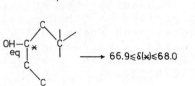 → $66.9 \leqslant \delta(*) \leqslant 68.0$

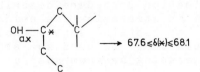 → $67.6 \leqslant \delta(*) \leqslant 68.1$

Beta Carbon Rules

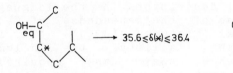 → $35.6 \leqslant \delta(*) \leqslant 36.4$

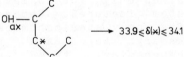 → $33.9 \leqslant \delta(*) \leqslant 34.1$

Gamma Carbon Rules

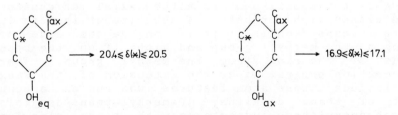

$20.4 \leqslant \delta(*) \leqslant 20.5$ $16.9 \leqslant \delta(*) \leqslant 17.1$

Figure 4. Sample rules constructed from decalins and hydroxy steroids with trans ring fusions. (*) identifies the carbon atom to which the shift is assigned; $\delta(*)$ is in ppm downfield from TMS.

equatorial orientation of the hydroxyl substituent. However, there is a difference between the pairs which could reflect the importance of the number of gamma carbons. The pair of beta carbon rules show the equatorial substitution results in a shift at lower fields than the axial substitution. The gamma rules shown illustrate an axial substituted hydroxy group will result in a gamma carbon shift at higher fields than that of an equatorial substituted hydroxy group.

Conclusion

Computer programs have been written to generate empirical C-13 NMR rules and to do structure elucidation using these rules. Rules generated on a set of paraffins and acyclic amines have successfully identified the C-13 NMR spectra of molecules not in the training set data. The form of the rule is suited for efficient structure elucidation using an algorithm which assembles substructures suggested by the rules as explanations of the observed shifts. The introduction of a limited set of stereochemical terms to the rule generation procedure demonstrated the feasibility of extending the method to more complicated systems.

Acknowledgements. This work was supported by the National Institutes of Health under grants 5R24 00612-07 and AM-17896-02 and by the Advanced Research Projects Agency under grant DAHC 15-73-C-0435.

Computer resources were provided by the SUMEX facility at Stanford University under National Institutes of Health grant RR-00785.

We are grateful to Jim McDonald, Bruce Buchanan, and Carl Djerassi for helpful discussions and William C. White for providing parts of the Meta-DENDRAL program code for use in this work.

Literature Cited

1. Stothers, J.B., "Carbon-13 NMR Spectroscopy," Academic Press, New York, N.Y. 1972.
2. Levy, G.C. and G.L. Nelson, "Carbon-13 Nuclear Magnetic Resonance for Organic Chemists," Wiley-Interscience, New York, N.Y. 1972.
3. Lindeman, L.P. and J.Q. Adams, Anal. Chem., (1971), 43,p. 1245.
4. Carhart,R. and C. Djerassi, J.Chem. Soc., Perkin Trans., (1973),2,p. 1753.
5. Eggert, H. and C. Djerassi, J. Amer. Chem. Soc. (1973),95,p. 3710.

6. Bremser, W., M. Klier, and E. Meyer, Org. Magn.
   Resonance, (1975),7,p. 97.
7. Jezi, B.A. and D.L. Dalrymple, Anal. Chem. (1975),
   47,p. 203.
8. Schwarzenbach, J., J. Meili, H. Konitzer, J.T.
   Clerc, Org. Magn. Resonance,(1976),8,p. 11.
9. Mitchell, T.M. and G.M. Schwenzer, to be published
10. Buchanan, B.G., D.H. Smith, W.C. White, R.J.
    Gritter, E.A. Feigenbaum, J. Lederberg, and C.
    Djerassi, J. Amer. Chem. Soc., (1976),98,
    p. 6168.
11. Buchanan, B.G., T.M. Mitchell, Proceedings of the
    Workshop on Pattern Directed Inference, Honolulu,
    Hawaii, (1977).
12. Carhart, R., D. Smith, H. Brown, and C. Djerassi,
    J. Amer. Chem. Soc., (1975),97,p. 5755.
13. Grover, S.H. and J.B. Stothers, Can. J. Chem.
    (1974),52,p. 870.
14. Eggert, H., C. VanAntwerp, N. Bhacca, and C.
    Djerassi, J. Org. Chem.,(1976),41,p. 71.

Computerized Structural Predictions from ^{13}C NMR Spectra

HENRY L. SURPRENANT and CHARLES N. REILLEY

Department of Chemistry, University of North Carolina, Chapel Hill, NC 27514

The wide chemical shift range and narrow line width of ^{13}C NMR spectra combine to yield an apparently limitless number of spectra.

For organic molecules, ^{13}C spectra are unique finger-printing tools and in an effort to understand ^{13}C chemical shift patterns, classes of compounds have been studied and spectral features correlated to structural changes within the class. These correlations are usually cast in the form of linear additivity parameters and have been largely successful for alkanes (1, 2) and monofunctional molecules, such as amines (3, 4) and alcohols (5).

An unassisted human cannot effectively consider the enormous number of possibilities that exist for molecular structures and ^{13}C spectra. A number of attempts have been made involving the use of computers to predict structures directly from ^{13}C spectra.

Burlingame, McPherson and Wilson (6) have correctly predicted the structures of seven acyclic saturated hydrocarbons up to empirical formula of $C_{20}H_{42}$. These compounds were not in the original data set used for the parameterization (2). Their procedure involved (a) nonredundantly generating codes for the structure of each geometrical isomer, (b) checking the code for the proper multiplicity (i.e., the correct number of primary, secondary, tertiary and quaternary carbons, informa-tion which can be obtained from an off-resonance decoupling experiment), (c) predicting the spectrum from the code using additivity parameters, and (d) comparing predicted and observed spectra, saving the best matches. Because their program could rapidly generate the structural codes, but was slow in computing the spectrum from that code, a multiplicity filter was employed to minimize computation time. An off-resonance decoupling experiment is particularly time consuming and its requirement is a major drawback of their approach. Although identification of seven structures is far from exhaustive proof of the utility, success in these cases is nevertheless

encouraging, particularly in light of the small number of
available spectra.

A more efficient approach was investigated by Carhart and
Djerassi (7) in the analysis of ^{13}C spectra of amines. Their
program (a) enumerated the possible substructures that could
account for the observed chemical shift of each line, (b) combined the substructures into larger fragments, "pruning"
illogical and mismatching fragments from the tree of possible
structures, and (c) generated the spectrum for each possible
structure and compared these with that of the unknown. All of
the 100 amine spectra studied, with the exception of six,
were unambiguously identified, and even in those six cases
the correct structure was always the best or second best match.
An interesting feature of their program was its ability to
predict structures from chemical shift information alone, not
requiring knowledge of the number of carbon atoms associated
with each peak.

The ^{13}C structural analysis problem was examined by Jezl
and Dalrymple (8) using a classical search system approach.
While polyfunctional and cyclic structures can readily be
handled within their search system, the ability to decipher
structures of compounds not present in their data file is, of
course, limited.

This paper summarizes a recent investigation (9) of the
uniqueness of ^{13}C spectra and how uncertainties in spectral
information affect the confidence in the predicted structures.
In addition, a method for handling functional groups as
derivatives of hydrocarbons has been developed, and a search
based system for hydrocarbon, amine, alcohol, and ether
molecules is demonstrated. Four FORTRAN programs have been
written to accomplish these goals. All programs were executed
on a Modular Computer Systems Model II/25 minicomputer. More
program information is available elsewhere (9).

Spectra Generation and Uniqueness Tests

Additivity Parameters. Linear additivity schemes have
been used in various forms in many scientific fields. The
application to ^{13}C NMR chemical shifts began with Grant and
Paul's (1) study of saturated acyclic hydrocarbons. Their
scheme showed significant deviations for highly branched
alkanes, and was modified by Lindeman and Adams (2) who also
expanded the observed data file. This parameterization scheme
has been used as a starting basis for including the effects of
the presence of functional groups such as amines (3, 4). The
standard error in the hydrocarbon parameterization was only
0.79 ppm indicating a high degree of reliability of the predicted spectra. This reliability and the large size of their
original data set were the primary reasons for the choice of
hydrocarbons for this study.

Structure Generation. The generation of all the nonredundant geometrical isomers for C_nH_{2n+2} hydrocarbons was
accomplished using program ISOMER. To help envision the
magnitude of the problem an enumeration of the number of
isomers vs. carbon number is listed in Table I. Because it
was not possible to check every structure code by hand, ISOMER
was assumed to execute properly because the total number of
isomers matched exactly with previous enumerations (10).

Table I. Number of Geometrical Isomers: C_nH_{2n+2}

n	No. of Isomers	n	No. of Isomers
1	1	11	159
2	1	12	355
3	1	13	802
4	2	14	1,858
5	3	15	4,347
6	5	16	10,359
7	9	17	24,894
8	18	18	60,523
9	46	19	148,284
10	75	20	366,319

Figure 1 gives a flow chart of the operation of ISOMER.
For a given number of carbon atoms, ISOMER first generates a
multiplicity (number of primary, secondary, tertiary and
quaternary carbons) and checks the validity of the multiplicity.
A line notation is generated for that multiplicity, and that
line notation stored on a disk file. The line notation is
then modified attempting to form new geometrical isomers of the
same multiplicity. Each nonredundant isomer is also stored
on the disk file. Because only four types of carbons occur
in a saturated hydrocarbon, only two bits are necessary to
encode each atom:

Code   Meaning

$00 = 0 = CH_3) = CH_3$ for first and last
$01 = 1 = CH_2$
$10 = 2 = CH($
$11 = 3 = C((

The parentheses are used to eliminate ambiguity of structures
around branches. A methyl group, which is a chain terminator,
ends the most recent chain started, not the first chain. This
notation is analogous to an arithmetic expression in which
the contents within the innermost parentheses are evaluated

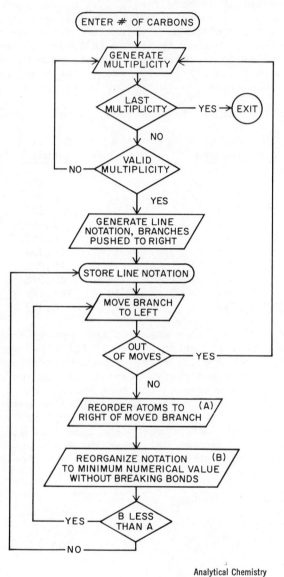

Analytical Chemistry

Figure 1.   Flowchart for program ISOMER (9)

prior to any outer parentheses. For example, 3-methyl-4-ethylheptane would be encoded as

$$CH_3CH_2CH_2CH(CH_2CH_3)CH(CH_3)CH_2CH_3$$

$$0 \quad 1 \quad 1 \quad 2 \quad 1 \quad 0 \quad 2 \quad 0 \quad 1 \quad 0$$

Another rule is that the arrangement of chains after a branch should give the lowest numerical code. For example, 3,4-dimethyl-4-ethylheptane would be encoded:

$$CH_3CH_2CH_2C((CH_3)CH_2CH_3)CH(CH_3)CH_2CH_3$$

$$0 \quad 1 \quad 1 \quad 3 \quad 0 \quad 1 \quad 0 \quad 2 \quad 0 \quad 1 \quad 0$$

and not as

$$CH_3CH_2CH_2C((CH(CH_3)CH_2CH_3)CH_2CH_3)CH_3$$

$$0 \quad 1 \quad 1 \quad 3 \quad 2 \quad 0 \quad 1 \quad 0 \quad 1 \quad 0 \quad 0$$

The structures are stored bit-compressed, 2 bits per atom, 8 atoms per 16-bit word.

Spectra Generation. The line notation structure codes from ISOMER are retrieved as well as program SPECGEN which then computes the expected ^{13}C spectra on the basis of Lindeman and Adams' additivity relationships (2). The flow chart for SPECGEN is given in Figure 2. Each line notation is read, unpacked, and processed to determine the number of bonds which separate each atom from every other atom in the molecule. On the basis of this information, a relatively straightforward algorithm can be constructed to add up proper constants according to the additivity parameter rules and thereby compute the predicted spectrum. A 1 ppm. bar graph showing the chemical shift distribution for the hydrocarbons C_2 through C_{16} is presented in Figure 3. The fine structure observed in the distribution is interesting and probably stems from regularities inherent in hydrocarbon molecular structures. While the chemical shifts for methyl groups fall in a different region from those for methylene groups, the extensive overlap prevents definitive differentiation between the two simply on the basis of their chemical shifts.

Uniqueness Tests. In this study, a ^{13}C spectrum was considered unique when, with each of its lines deviating by a given amount from the predicted value, the proper structure was still the best matching and non-unique otherwise. We found that, in almost every instance when the proper structure was not the best matching, it was the second best matching. The matching procedure was simple when the number of lines in

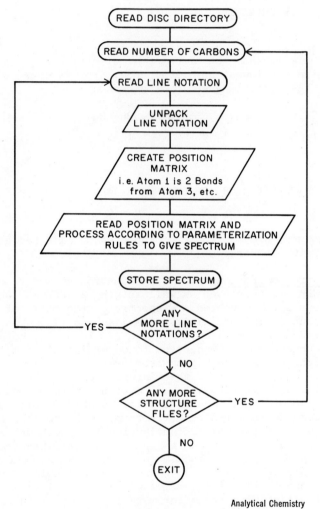

Figure 2. Flowchart for program SPECGEN (9)

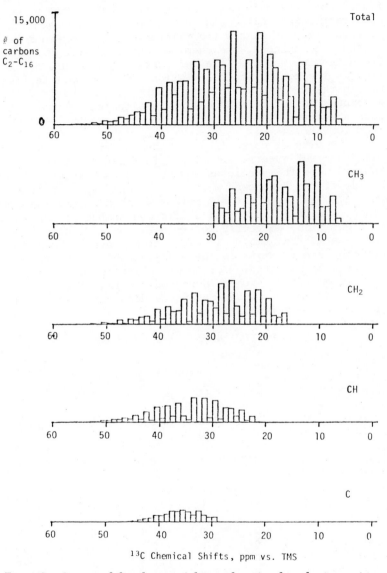

Figure 3.   Computed distribution of the number of carbons having various ¹³C chemical shifts, C₂–C₁₆

the reference and unknown spectra were equal. The chemical shifts of both the reference and the unknown spectra were ordered from low to high, and the sum of the squares of the deviations computed. When the reference spectrum had more lines than the unknown, the chemical shifts were again ordered, but to determine which of the reference lines should be omitted in the comparison calculation involved more testing. To obtain the minimum sum of the squares of the deviations each combination of reference lines was tested against the unknown spectrum. These functions were implemented by the program COMPARE.

The results of comparisons of the spectral file with normal "unknown" spectra are listed in Table II. The "unknowns" were the predicted spectra for each isomer but where each line of the "unknown" spectrum was shifted by a given amount, Δ. The term "normal" refers to those proton-decoupled ^{13}C spectra with all chemical shifts and number of carbons (via intensities) correctly determined. As indicated in Table II until the number of carbon atoms becomes large or a significant error exists in the chemical shift, the correct structure was predicted. For large molecules, C_{13} and above, the number of wrong matches increased. Here, it may be necessary to perform an off-resonance decoupling experiment to determine the multiplicities.

Table II. Normal ^{13}C Spectra

	% Wrong Matches					
No. of Carbons	9	10	11	12	13	14
No. of Isomers	35	75	159	355	802	1858
0.11[a]	0	0	0	0	0	0
0.22	0	0	0	0	0	0
0.33	0	0	0	0	0	0
0.44	0	0	0	0	0	0
0.55	0	0	0	0	0	0
0.66	0	0	0	0	0	0.1
0.77	0	0	0	0	0.2	0.5
0.88	0	0	0	0	0.5	0.9
0.99	0	0	0	0.3	1.4	2.1
1.10	0	0	0	0.6	2.5	4.6
1.21	0	0	1.9	1.7	4.5	11.1
1.32	0	1.3	2.5	2.5	9.7	20.1
1.43	0	1.3	4.4	5.9	15.8	32.4
1.54	0	2.6	6.9	11.0	26.7	44.8
1.65	0	2.6	8.8	20.6	37.9	56.6

[a]Vertical column shows Δ, deviation of each line, (ppm).

Excellent confidence can be placed in the predicted structure, even for C_{18}-C_{20} molecules, if the multiplicity is known. This is evidenced by the strikingly low percentage of wrong matches, even for spectra whose chemical shifts are quite deviant; see Table III.

Table III. ^{13}C Spectra with Known Multiplicities

No. of	% Wrong Matches		
Carbons	12	13	14
No. of			
Isomers	355	802	1858
0.11[a]	0	0	0
0.22	0	0	0
0.33	0	0	0
0.44	0	0	0
0.55	0	0	0
0.66	0	0	0
0.77	0	0	0.05
0.88	0	0	0.05
0.99	0	0	0.05
1.10	0	0	0.05
1.21	0	0	0.11
1.32	0	0.12	0.16
1.43	0	0.12	0.16
1.54	0	0.12	0.22
1.65	0	0.12	0.48

[a]Vertical column shows Δ, deviation of each line, ppm.

The values given in Tables II and III assume, however, that intensity information was sufficiently accurate to permit correct determination of the number of carbons under each peak, and that each peak is observable. Because of the variability of the nuclear Overhauser enhancement (NOE) and wide range of relaxation times, inaccurate area measurements are the rule and not the exception with typical ^{13}C spectra. Indeed, carbons with no attached protons are often not observed. Two additional uniqueness studies were carried out to study the influence of inaccurate intensity information.

As indicated in Table IV where it is assumed that quaternary carbons were not observed, confidence in prediction drops considerably. Even more dramatic is the drop in confidence if intensity information is sufficiently poor so that the number of carbons for a given peak cannot be assigned and thus lines for equivalent carbons cannot be distinguished from other lines, Table V. These results point to potential problems arising from the manner in which spectra are obtained. High pulse repetition rates without either gated decoupling or purposeful addition of relaxation reagents often reduce the intensity of quaternary carbons and distort areas for other carbons.

Table IV. ^{13}C Spectra with Unobserved Quaternary Carbons

No. of Carbons	% Wrong Matches						
	8	9	10	11	12	13	14
No. of Isomers	18	35	75	159	355	802	1858
0.11[a]	0	0	0	0	0	0	0
0.22	0	0	0	0	0	0	0
0.33	0	0	0	0	0	0	0
0.44	0	0	0	0	0	0	0.2
0.55	0	0	0	0	0.5	0.2	0.5
0.66	0	0	0	0	0.5	0.4	1.5
0.77	0	0	0	0	0.5	1.1	4.1
0.88	0	0	0	1.1	0.9	3.4	7.1
0.99	0	0	0	3.2	5.5	10.4	17.8
1.10	0	0	0	5.4	10.5	19.0	30.3
1.21	0	0	0	5.4	15.0	29.6	44.7
1.32	0	5.9	5.3	12.9	22.7	42.8	57.5
1.43	0	5.9	7.9	22.6	30.5	52.1	67.8
1.54	0	11.8	15.8	26.9	39.5	61.3	75.9
1.65	0	17.6	21.0	31.2	52.7	69.8	81.6
Number of isomers containing quaternary carbon(s)							
	7	17	38	93	220	537	1306

[a]Vertical column shows Δ, deviation of each line, ppm.

The conclusions are obvious. In order to identify large molecules nearly unambiguously, ^{13}C spectra must be obtained without saturating quaternary resonances, with minimum NOE effects, and perhaps with off-resonance decoupling. It is also advisable to use the same temperature and solvent conditions utilized in obtaining the data base employed in the parameterization.

Hydrocarbon Based ^{13}C Search System

Program MATCH is a hydrocarbon based ^{13}C search system designed to handle saturated acyclic hydrocarbons and related monofunctional amines, alcohols, and ethers. The number of ^{13}C peaks observed, the type of functional group search desired, and the observed chemical shifts are the inputs to the program. Correct intensity information is implied by inputting the proper number of resonances. MATCH outputs the ten best structures and the assignment of the spectrum. In Figure 4 the outline of the program is depicted.

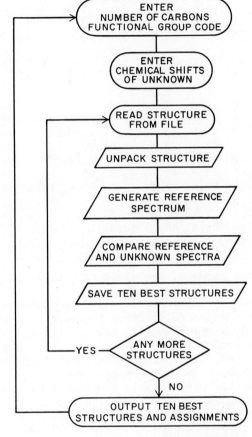

Analytical Chemistry

Figure 4.   Flowchart for program MATCH (9)

Table V. Equivalent Carbons Treated as One Carbon

No. of Carbons	% Wrong Matches							
	7	8	9	10	11	12	13	14
No. of Isomers	9	18	35	75	159	355	802	1858
0.11[a]	0	0	0	5.8	2.0	4.8	3.0	1.8
0.22	0	0	0	7.2	3.4	6.0	4.1	2.6
0.33	0	0	9.4	7.2	4.8	6.9	4.6	4.2
0.44	0	0	9.4	8.7	6.1	7.9	6.2	7.3
0.55	0	5.9	9.4	13.0	7.2	9.1	8.0	9.1
0.66	0	5.9	9.4	13.0	9.5	12.1	12.2	15.2
0.77	0	17.6	15.6	15.9	10.9	16.0	18.8	23.8
0.88	0	17.6	15.6	20.3	12.9	20.5	29.2	36.8
0.99	0	23.5	18.8	24.6	20.4	27.5	38.3	49.2
1.10	0	23.5	21.9	26.1	23.8	34.7	49.2	60.0
1.21	12.5	23.5	21.9	31.9	30.6	43.2	58.3	67.9
1.32	12.5	23.5	21.9	39.1	37.4	50.5	65.4	75.8
1.43	12.5	29.4	25.0	44.9	44.9	56.5	71.7	81.6
1.54	12.5	29.4	31.3	49.3	53.1	63.4	77.2	86.0
1.65	25.0	35.3	46.9	50.7	57.1	69.2	80.9	88.8
Number of isomers containing equivalent carbons	8	17	32	69	147	331	760	1779

[a]Vertical column shows Δ, deviation of each line, ppm.

MATCH reads the file of encoded hydrocarbon structures and computes the expected spectrum for each structure, and then compares each spectrum with that of the unknown. The storage of the compressed structure codes, but not the associated spectra, saves significant amounts of storage without undue degradation of searching speed. Storage of the expected hydrocarbon spectra would double the searching speed for hydrocarbons, but would not change the speed for functional group searches.

The hydrocarbons listed in Table VI, in addition to the 69 hydrocarbons utilized in forming the parameters data (2) were correctly identified by MATCH using their chemical shifts and intensity information.

To handle functional groups, MATCH simply searches that structure file for hydrocarbons which is one carbon larger than the molecule of interest. Nonequivalent carbons in the structures are then determined and the functional group replaces one of the nonequivalent carbons to obtain one new molecule. Because a given hydrocarbon isomer can have more than one nonequivalent carbon, more than one isomer of the functional groups substituted molecule is obtained from each hydrocarbon. For example, all the monofunctional C_{18} alcohol isomeric structures can be derived from the C_{19} hydrocarbon

Table VI. Hydrocarbons Identified by Program MATCH

Unknown Compound	Number of Carbons	Number of Possible Isomers
n-Decane	C_{10}	75
2,7-Dimethyloctane	C_{10}	75
4-n-Propylheptane	C_{10}	75
2,4,6-Trimethylheptane	C_{10}	75
2,8-Dimethylnonane	C_{11}	159
2,5,8-Trimethylnonane	C_{12}	355
2,9-Dimethyldecane	C_{12}	355
n-Hexadecane	C_{16}	10,359
2,5,8,11-Tetramethyldodecane	C_{16}	10,359

file, by replacing, one at a time, each nonequivalent methyl group with a hydroxyl. Similarly, for ethers nonequivalent methylenes are replaced with an oxygen. Some of the functional groups that can be handled in this manner are shown in Table VII. Once the isomeric structures are obtained, the spectra can quickly be generated using the additivity parameters for that functional group (4, 5).

Table VII. Extension of Hydrocarbon Scheme
to Cover Functional Groups

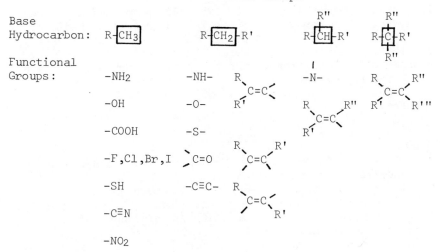

The replacement of a carbon with a functional group increases the number of possible geometrical isomers manyfold. Table VIII lists the enumeration of the number of monofunctional hydrocarbon isomers as a function of the number of carbons and type of functional group. While 10,359 isomers exist for $C_{16}H_{34}$, the corresponding number of amine isomers for $C_{15}H_{33}N$

Table VIII.   Number of Isomers Having
Potentially Nonequivalent Spectra

(Monofunctional Groups)

No. of Carbons	No. of Hydrocarbon Isomers	No. of nonequivalent carbons			
		CH_3	CH_2	CH	C
2	1	1	0	0	0
3	1	1	1	0	0
4	2	2	1	1	0
5	3	4	3	1	1
6	5	8	6	3	1
7	9	17	15	7	3
8	18	39	33	17	7
9	35	89	82	40	18
10	75	211	194	102	42
11	159	507	482	249	109
12	355	1238	1188	631	269
13	802	3057	2988	1594	691
14	1858	7639	7527	4073	1759
15	4347	19241	19181	10442	4542
16	10359	48863	49056	26974	11732

is 48,863 + 49,056 + 26,974 or 124,893. With this increase
in number of isomers, a decrease in confidence in predicted
structures is expected. Approximately 50 amine spectra (3)
and 10 alcohol spectra (5) were all correctly identified by
program MATCH operating in the functional group mode. For a
C_{15} amine approximately 1/2 hour was required for the identification, searching through 124,893 amine structures and
spectra.

Abstract

     On the basis of computer generated isomeric structures
for acyclic saturated hydrocarbons and Lindeman and Adams'
parameter values through γ-carbons, a number of questions
concerning ^{13}C NMR spectra were addressed. The principle
features included determination of the uniqueness of the
spectra and the influence of uncertainties in the spectral
information which can arise when multiplicities of individual
carbons is unknown (i.e. off-resonance data not available),
when quaternary carbon signals are not observed (i.e. because
of saturation), and when equivalent carbons are treated as
one carbon. Methods of handling the presence of monofunctional
groups and utilization of the above information to form the
basis of a search system for hydrocarbon, amine, alcohol, and
ether molecules are also considered.

Acknowledgements

The authors wish to thank the National Science Foundation and the National Institutes of Health for their financial support of this study.

Literature Cited

1. Grant, D. M. and E. G. Paul, J. Am. Chem. Soc. (1964) 86, 2984.
2. Lindeman, L. P. and J. Q. Adams, Anal. Chem. (1971) 43, 1245.
3. Eggert H. and C. Djerassi, J. Am. Chem. Soc. (1973) 95, 3710.
4. Sarneski, J. E., H. L. Surprenant, F. K. Molen, and C. N. Reilley, Anal. Chem. (1975) 47, 2116.
5. Roberts, J. D., F. J. Weigert, J. I. Kroschwitz, and H. J. Reich, J. Am. Chem. Soc. (1970) 92, 1338.
6. Burlingame, A. L., R. V. McPherson, and D. M. Wilson, Proc. Nat. Acad. Sci., USA (1973) 70, 3419.
7. Carhart, R. E. and C. Djerassi, J. Chem. Soc., Perkin Trans. 2 (1973) 1753.
8. Jezl, B. A. and D. L. Dalrymple, Anal. Chem. (1975) 47, 203.
9. Surprenant, H. L and C. N. Reilley, Anal. Chem. in press.
10. Lederberg, J., G. L. Sutherland, B. G. Buchanan, E. A. Feigenbaum, A. V. Robertson, A. M. Duffield, and C. Djerassi, J. Am. Chem. Soc. (1969) 91, 2973.

7

Interactive Structure Elucidation

C. A. SHELLEY, H. B. WOODRUFF, C. R. SNELLING, and M. E. MUNK

Department of Chemistry, Arizona State University, Tempe, AZ 85281

The topic of computer-assisted structure elucidation cuts across disciplinary boundaries as well as the traditional boundaries within the discipline of chemistry itself. As a result, scientists with varied backgrounds and interests have been attracted to it. We entered the area through the door marked "natural products chemists." It was our own work in structure elucidation (1-8) that led us to the computer and the belief that the process as practiced by the natural products chemist is amenable to computer modeling. It is quite evident that our efforts bear the imprint of our background.

While it is true that no two natural products chemists practice the science and art of structure elucidation in exactly the same way, certain common features may be discerned (Figure 1). Three integral components of the process are:

1. the reduction of chemical and physical data to their structural implications;
2. the familiar partial structure, an expression comprised of known structural fragments and unaccounted-for atoms that summarizes the status of the structure problem at any given stage;
3. the design of new experiments, guided by the partial structure, or some or all of the molecular structures compatible with it.

The final solution of the problem may be described as the cyclic process that leads to the reduction of the number of structural fragments and atoms in the partial structure to one.

Description of CASE

In developing a computer model of the structure

elucidation process as we envisioned it, we first
focussed our attention on two of its major components:

1. the expansion of a partial structure to all
    molecular structures consistent with it and
    any other information available to the chem-
    ist, and
2. the reduction of chemical and spectroscopic
    data to their structural implications.

The current status of CASE, our acronym for
computer-assisted structure elucidation, is summarized
in Figure 2. CASE is a network of computer programs
designed to accelerate and make more reliable the
entire process of structure elucidation. The system
is highly interactive and is continually evolving.

The task of reducing chemical and spectroscopic
data to structural information is presently shared by
the chemist and the computer. An infrared interpreter
designed specifically for application to multifunc-
tionalized molecules is at an advanced stage of development and fully operational (9,10). If the interpreter is to be of value to the natural products
chemist in solving actual structure elucidation problems, several criteria must be met. The program must
be able to make decisions concerning the presence or
absence of a large number of functional groups. It is
insufficient simply to distinguish esters from non-
esters. Rather one should be able to make a more
specific distinction (e.g., saturated esters vs. unsaturated esters vs. lactones). In addition, the
program must be able to interpret the relatively
complex spectra of compounds found in nature. Our
program tests for the presence or absence of 169
chemical functionalities. It has been tested on over
500 spectra of varying complexity with a high degree
of success.

It is an artificial intelligence program that
attempts to parallel the chemist's reasoning in interpreting an infrared spectrum as much as possible.
The chemist uses an empirical approach to interpret
infrared spectra. A set of guidelines for interpreting infrared spectra is determined. These guidelines may result from observation of a sufficient
number of spectra and/or from reading textbooks and
learning from the observations of others. An initial
set of guidelines for identifying saturated carboxylic
acids might be to look for a broad, medium to strong
peak centered around 3000 cm^{-1}, a strong carbonyl peak
near 1715 cm^{-1}, and a broad, medium intensity peak in
the vicinity of 920 cm^{-1}. Any compounds with spectra
which followed these guidelines would be interpreted

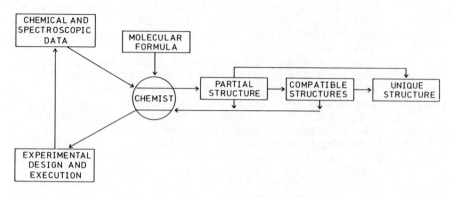

Figure 1. "Manual" structure elucidation

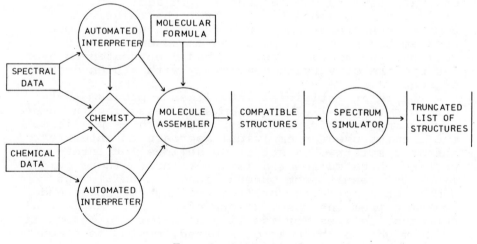

Figure 2. CASE network

as containing a carboxylic acid functionality. Similar initial sets of guidelines must be established for all other functionalities. Next the program is tested on some infrared spectra. Anytime the program makes an erroneous interpretation, it is necessary to alter the guidelines to correct the mistake. As long as a chemist can do a better job of interpreting a spectrum than the program, then the program can be altered by using this new information so that it will do a better job of interpretation in the future. This ability to alter the program to correct mistakes is a major advantage of artificial intelligence programming. As with other components of CASE, the infrared interpreter is continually evolving.

The computer-derived analysis of the infrared spectrum can be reviewed by the chemist prior to automatic encoding, or the chemist can be bypassed, with the automatically encoded functional group information being sent directly to the molecule assembler. The development of programs for the automated interpretation of other spectroscopic information is at an earlier stage. A preliminary investigation using pattern recognition techniques to aid in the interpretation of ^{13}C-NMR spectra has yielded promising results. At the present time, much of the remaining interpretation of spectral data is chemist-derived.

CASE also incorporates programs for the automated interpretation of chemical information. As an example, CASE accepts the number of moles of periodate consumed by a compound and uses the information to constrain the molecule assembler. Thus, only molecules consistent with the periodate information are assembled.

The molecule assembler accepts both the computer-derived and user-derived structural information, and, given the molecular formula, constructs all structural isomers compatible with the input. A nonredundant listing of the compatible molecules is presented to the chemist in the conventional structural language.

The list of compatible molecules usually can be further truncated by comparing certain predicted spectroscopic properties for each molecule constructed with the observed spectroscopic properties of the unknown. The tasks of predicting, comparing, and ranking are assigned to the spectrum simulator. These programs are also at an early stage of development. One operational component called PEAK will be illustrated later. Given a molecular structure, PEAK predicts the number of signals expected in the ^{13}C-NMR spectrum.

The molecule assembler (Figure 3) is unique in approach and in its simplest form was designed to expand exhaustively the conventional partial structure into all structural isomers consistent with it. The algorithm on which the assembler is based is recursive. By concentrating more on partial structure expansion than on molecular formula expansion, greater compactness and efficiency were achieved.

From the start it was recognized that a broadly applicable molecule assembler must be capable of utilizing more information than just the conventional partial structure, because the chemist generally has more information than can be expressed by the conventional partial structure.

Thus, in Figure 3 the term "partial structure" is used in a broader context. The partial structure includes the molecular formula and computer-derived and/ or chemist-derived structural fragments. Atoms in these fragments must not duplicate one another; that is, the fragments must be nonoverlapping. The partial structure also includes supplementary information that cannot be expressed in terms of nonoverlapping structural fragments.

Communication with CASE is achieved by means that mimic the natural language of the chemist. The molecular formula is input in standard format. The versatile linear code designed for structural fragment input is illustrated in Figure 4. Thus, lines 1, 3, 5, 7 and 9 represent respectively, a 1-hydroxyethyl group with a single residual valence at the 1-position, a carbonyl group with doubly deficient carbon atom, a carbonyl group joined to oxygen with valence deficiencies at carbon and single bonded oxygen, a carboxyl group and a cyclohexane ring with each carbon atom doubly valence deficient. In the latter example, note that ring designation is achieved by labeling one carbon atom with a number (1 in this case) and forming a bond between a carbon five atoms removed and the label.

The linear code may be further elaborated by adding atom and fragment tags to describe the local environment of atoms in a structural fragment without concern for overlapping atoms. For example, the valence deficient carbon atom of the 1-hydroxyethyl group is required to bond to a methine carbon by the addition of the atom tag <CH1> (line 2). Since the information is provided by means of the tag, the chemist need not be concerned whether that methine carbon duplicates a similar group in another structural fragment.

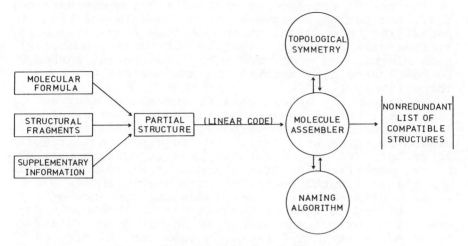

Figure 3. Molecule assembler

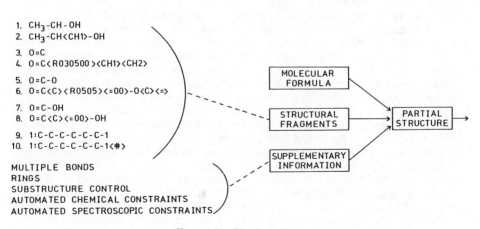

Figure 4. Constraints

Statement 4 restricts the environment of the carbonyl carbon atom with three atom tags: <R030500> prohibits its presence as part of a 3-5 membered ring, i.e., the carbonyl group is unstrained (R for ring, 03 for minimum ring size, 05 for maximum ring size, 0 for minimum number of such rings, 0 for maximum number of such rings), and <CH1> and <CH2> require the carbonyl to be flanked by a methine and a methylene group, respectively.

Statement 6 specifies a five-membered lactone with unsaturation conjugated to the alcohol oxygen. The environment of the carbonyl carbon atom is disclosed by three atom tags: <C> requires that it join to carbon, <R0505> designates that it must be part of a five-membered ring (the absence of minimum and maximum values for the number of such groups infers a minimum of 1) and <=00> precludes the presence of α,β-unsaturation to that carbon atom (= is the double bond symbol, 0 for minimum number of double bonds, 0 for maximum number). The two atom tags for the alcohol oxygen require that it bond to carbon (<C>) and the presence of α,β-unsaturation (<=>).

Statement 8 represents a carboxylic acid function that cannot bear α,β-unsaturation. The symbol <#> is a fragment tag that instructs the program to forbid the formation of internal bonds in a fragment. Thus, because of statement 10, no molecular structures containing multiple bonds or bridges in the cyclohexane ring will be assembled.

Some structural information is not specific to atoms or fragments and must be treated separately. Thus, Multiple Bonds designates the number, either exact or a range, and kinds of multiple bonds allowed; Rings permits the same control over rings. Substructure Control can be used to require the presence or absence of any specified structural fragment using the same linear code described. Automated Chemical and Spectroscopic Constraints instruct the molecule assembler to generate only those structures consistent with the applied constraints. For example, the applied constraint may be the number of moles of periodate that an unknown compound consumes or the number of signals in the ^{13}C-NMR spectrum of the unknown. These applications will be illustrated later.

The information contained in "partial structure" (Figure 3) is used in a way that maximizes the efficiency of the molecule assembler. The goal is to preclude the assembly of an invalid molecule rather than reject it after assembly. Thus, in most cases, atom and fragment tags, and supplementary information constrain the assembly process. In other cases, retrospective searching is required.

The molecule assembler is constrained in other ways as well.

1. A user-defined library of highly strained or chemically unstable moeities constrains the assembly of molecules containing these structural features, e.g., a cyclopropanone ring.

2. The assembly of duplicate structures is minimized. A number of techniques are incorporated into the program to eliminate duplicates prospectively. The most important technique involves the perception of topological symmetry (11) at each step of the assembly process. In this way, only one member of a group of topologically equivalent valence-deficient atoms is permitted to initiate bond formation. In spite of the heuristics used, the total exclusion of duplicates cannot always be assured. Those duplicates still formed can be eliminated retrospectively. A newly designed and highly efficient canonical naming algorithm performs this remaining task. The algorithm also recognizes resonance forms and includes only one member on the list of valid structures.

Applications

In the discussion of applications, our purpose is to provide an overview of the system network, its interactive nature, and its scope. CASE was originally developed on real world problems under study in our own laboratory and in a couple of other natural products laboratories. In more recent years "simulated" real world problems, that is, problems taken from the chemical literature, have also played an important role. Such problems, although admittedly somewhat contrived, provided the necessary breadth or depth exactly when needed in the program development and testing. Specific kinds of problem situations are difficult to produce on demand with slower moving real world structure problems.

Faulkner et al. (12) recently isolated a halogenated monoterpene from a sea hare and reported the structure (A). The assignment was made in part on the chemical and spectroscopic data reported in the paper and in part by analogy to a related known compound. The structural information derived from the chemical and spectroscopic data is shown in the I/O printout of

the initial run of the problem (Figure 5). The molecular formula is input as $C_{10}H_{13}B_2X_3$, where X, a
standard halogen symbol, is Cl in this problem, and B
is user-defined as Br. Since B is not a standard molecular formula symbol, the computer asks the user to
specify its valence. Next the nonoverlapping fragments are listed. In order they are: an allyl
chloride fragment with all carbon atoms having residual valence, a vinyl bromide unit, a bromomethylene
unit, and two CH_3 groups each attached to a quaternary
carbon. The structure constraining device, Substructure Control, is used to limit the number of CH_3
groups to the known value of 2.

At this point we are ready to start the molecule
assembler. Since we have no estimate on the total
number of structural isomers consistent with this
input, we set some arbitrary limit on molecule assembly, 20 in this case, to make certain that the problem
strategy is indeed sound before giving free rein to
the molecule assembler. The first 20 structures were
examined and we found that we could easily constrain
the molecule assembler with 2 additional pieces of
information because we saw a conjugated diene in some
structures and a gem-dimethyl group in some structures.
One of the authors of the paper, Dr. Ireland, indicated the availability of evidence, not in the paper,
to exclude a conjugated diene. In addition, the
geminal methyl groups should have been excluded by us
based on the published PMR information.

These two constraints were added by calling
Substructure Control (Figure 6). Sixteen structures
were generated. Of the 16, 8 had geminal chlorine
atoms. Ireland believed these structures to be unlikely on the basis of indirect chemical evidence.
The remaining 8 candidates could be pruned to 4 structures on the basis of some of the mass spectroscopic
data. The structure assigned (A) is one of the four.

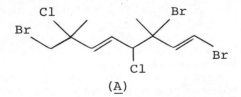

(A)

In our own structure work on the antibiotic
actinobolin (5), which is a compound unrelated in
structural type to known antibiotics, an earlier
version of CASE was used. The structure study was
initiated by an examination of a degradation product,

```
TITLE: FAULKNER   JOC,41,2461(1976).
ENTER THE MAXIMUM NUMBER OF STRUCTURES TO BE GENERATED: 20
MOLECULAR FORMULA: C10H13B2X3
WHAT IS THE VALENCE OF A B ATOM?: 1
FRAGMENT(S): CH=CH-CH-X
FRAGMENT(S): CH=CH-B
FRAGMENT(S): CH2-B
FRAGMENT(S): CH3<CHO> CH3<CHO>
FRAGMENT(S): ;
CONSTRAINT(S): SUBSTR
FRAGMENT(S): CH3 ;
MINIMUM: 2
MAXIMUM: 2
CONSTRAINT(S): ;
COMMAND: GENERATE


      20 STRUCTURES GENERATED
```

Figure 5

```
            TITLE: FAULKNER   JOC,41,2461(1976).
            MOLECULAR FORMULA: C10H13B2X3
            WHAT IS THE VALENCE OF A B ATOM?: 1
            FRAGMENT(S): CH=CH-CH-X
            FRAGMENT(S): CH=CH-B
            FRAGMENT(S): CH2-B
            FRAGMENT(S): CH3<CHO> CH3<CHO>
            FRAGMENT(S): ;
            CONSTRAINT(S): SUBSTR
            FRAGMENT(S): CH3 ;
            MINIMUM: 2
            MAXIMUM: 2
            CONSTRAINT(S): SUBSTR
            FRAGMENT(S): C=C-C=C ;
            MINIMUM: 0
            MAXIMUM: 0
            CONSTRAINT(S): SUBSTR
            FRAGMENT(S): CH3-C-CH3 ;
            MINIMUM: 0
            MAXIMUM: 0
            CONSTRAINT(S): ;
            COMMAND: GENERATE


               16 STRUCTURES GENERATED
```

Figure 6

actinobolamine (B̲), containing 9 of the original 13
carbon atoms (2̲). That problem has been rerun on the
current version of CASE.

The IR interpreter program reports the presence
of alcohol and ketone with high confidence levels, 4
and 3 (Figure 7). Subclass information, with confidence levels of 2, is of insufficient certainty for
use by the molecule assembler. The same is true of
other major classes of functional groups listed. Note
that the program considers only functional groups
consistent with the molecular formula. In addition,
atoms of functional groups receiving a confidence
level of 4 are automatically subtracted from the
molecular formula and are not considered further.

The I/O printout shown in Figure 8 includes
unstrained ketone carbonyl, i̲.e̲., not part of a 3-5
membered ring, flanked by a C̄ bearing four readily
exchangeable protons, a secondary hydroxyl group, a
secondary amine attached to two different methine
carbons, and a 1-hydroxyethyl group. The Multiple
Bond Constraints DOUBLE and TRIPLE preclude the
generation of double and triple bonds. Substructure
Control restricts assembly of ketal and aminal
linkages, and also the number of CH₃ groups to one.
The consumption of two moles of periodate by actinobolamine is important structural information, the
significance of which is automatically considered by
calling PERIODATE and giving the molar uptake. This
information constrains the molecule assembler itself;
the search for compatibility is not done retrospectively.

CASE produced five structures consistent with the
available evidence (Figure 9). Armed with the assur-
ance that no valid structure had been overlooked, an
examination of these five structures provided invalu-
able guidance in the design of the minimum number of
experiments to assign the structure of actinobolamine
correctly. All of the experiments were spectroscopic
in nature and led to the correct structure (B̲).

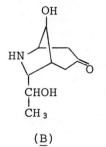

(B̲)

$$C13H20N2O6 \xrightarrow{H3O^{+}} C9H15NO3$$

ACTINOBOLIN   ACTINOBOLAMINE

IR INTERPRETER OUTPUT

EACH CLASS HAS A CONFIDENCE LEVEL OF 0-4.
        4 - DEFINITELY PRESENT
        3 - HIGH PROBABILITY
        2 - MEDIUM PROBABILITY
        1 - LOW PROBABILITY
        0 - DEFINITELY ABSENT

            CLASS   SUBCLASSES

1. ALCOHOL-------------- 4   2. PRIMARY------------ 2   3. 2-A,B-------------- 2
    ---------------------   4. 3-A,B-A',B'-------- 2   5. SEC. IN RING------- 2

6. KETONE-------------- 3   7. SATURATED---------- 2

8. LACTAM------------- 2   9. 5 MEMBER W/O NH--- 2

10. CARBAMATE---------- 2   11. PRIMARY------------ 2 12. SECONDARY---------- 2
     --------------------   13. TERTIARY---------- 1

14. AMINE------------- 2   15. SECONDARY--------- 2 16. TERTIARY----------- 1

17. C=C(NON-AROMATIC) 2   18. CHR=CR2----------- 1

19. METHYL------------ 2   20. GEM DIMETHYL------ 1

21. NITRO GROUP------- 2   22. SATURATED--------- 2

23. PYRROLE----------- 2

24. ETHER------------- 1   25. SATURATED--------- 1 26. UNSATURATED------- 1

27. ACETAL----------- 1

28. KETAL----------- 1

Figure 7

```
TITLE: ACTINOBOLAMINE
MOLECULAR FORMULA: C9H15N1O3
FRAGMENT(S): O=C<RO30500><CH222>
FRAGMENT(S): OH<CH111>
FRAGMENT(S): NH<CH122>
FRAGMENT(S): CH3-CH-OH
FRAGMENT(S): ;
CONSTRAINT(S): DOUBLE    O    O
CONSTRAINT(S): TRIPLE    O    O
CONSTRAINT(S): SUBSTR
FRAGMENT(S): O-C-O ;
MINIMUM: O
MAXIMUM: O
CONSTRAINT(S): SUBSTR
FRAGMENT(S): O-C-N ;
MINIMUM: O
MAXIMUM: O
CONSTRAINT(S): SUBSTR
FRAGMENT(S): CH3 ;
MINIMUM: 1
MAXIMUM: 1
CONSTRAINT(S): PERIODATE    2
CONSTRAINT(S): ;
COMMAND: GENERATE
```

Figure 8   5 STRUCTURES GENERATED

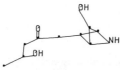
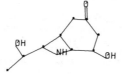
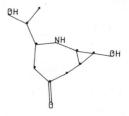
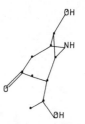
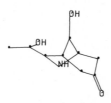

Figure 9.   Program CASE-draw, Arizona State University, actinobolamine

Coronatine is a toxin produced by a microorganism of the Pseudomonas genus. Its structure was reported early this year (13). A degradation product, coronafacic acid, played a key role, and although some chemical and spectroscopic data and their structural significance are reported in the paper, the final determination of the degradation product was by x-ray. We decided to see how close to the actual structure the reported chemical and spectroscopic information would have taken the authors. The computer input is shown in Figure 10.

Coronafacic acid is $C_{12}H_{16}O_3$. It contains a cyclopentanone ring with three readily exchangeable hydrogen atoms, an unstrained α,β-unsaturated carboxylic acid moeity and an ethyl group that is not part of a propyl group. There are no additional multiple bonds and only a single methyl group. CASE assembled 88 structures, thus, the chemical and spectroscopic evidence brought the authors to within 88 structures of the correct one.

One last simple, but informative example illustrates one application of the spectrum simulator (Figure 11). The monoterpene cineole, $C_{10}H_{18}O$, was recently examined by ^{13}C-NMR. The off-resonance and broad-band proton decoupled spectra reveal quaternary carbon bearing ether oxygen, at least one methine carbon, two methylene carbons and two methyl carbons, and no unsaturated carbons. The ^{13}C-NMR evidence is compatible with 458 structural isomers according to CASE.

If PEAK is called (Figure 12), the number of ^{13}C-NMR signals expected for each of the 458 compounds is predicted. Those structures not conforming to the observed number, 7 in this case, are rejected. In this way the list of 458 structures is pruned to 38. Of the 38 structures, only 5 conform to the isoprene rule.

Peak prediction is based on molecular topology, but the determination of class equivalence in this case considers only neighboring atoms no more than three bonds removed. Since a perfect match between prediction and observation cannot be expected for each and every structure examined by PEAK, the pruning step of PEAK can compare the actual number of observed signals to a range of predicted values, generally the actual number plus or minus one.

Thus, if PEAK is set at 7 with a range of plus or minus one, the list of 458 structures is reduced to 144. Of the 144, only 19 comply with the isoprene rule.

```
TITLE: CORONAFACIC ACID
MOLECULAR FORMULA: C12H1603
FRAGMENT(S): 1:C(=O)-CH<HOO>-C-C-CH2-1<#>
FRAGMENT(S): CH<RO30500><HOO>=C<HOO>-C(=O)-OH
FRAGMENT(S): CH3-CH2<CH2OO>
FRAGMENT(S): ;
CONSTRAINT(S): DOUBLE   0   0
CONSTRAINT(S): SUBSTR
FRAGMENT(S): CH3 ;
MINIMUM: 1
MAXIMUM: 1
CONSTRAINT(S): ;
COMMAND: GENERATE


      88 STRUCTURES GENERATED
```

Figure 10

```
              TITLE: CINEOLE
              MOLECULAR FORMULA: C10H18O1
              FRAGMENT(S): C<HOO>-O-C<HOO>
              FRAGMENT(S): CH<HOO>
              FRAGMENT(S): CH2<HOO> CH2<HOO>
              FRAGMENT(S): CH3 CH3 ;
              CONSTRAINT(S): DOUBLE   0   0
              CONSTRAINT(S): TRIPLE   0   0
              CONSTRAINT(S): ;
              COMMAND: GENERATE
```

Figure 11   458 STRUCTURES GENERATED

```
              TITLE: CINEOLE
              MOLECULAR FORMULA: C10H18O1
              FRAGMENT(S): C<HOO>-O-C<HOO>
              FRAGMENT(S): CH<HOO>
              FRAGMENT(S): CH2<HOO> CH2<HOO>
              FRAGMENT(S): CH3 CH3 ;
              CONSTRAINT(S): DOUBLE   0   0
              CONSTRAINT(S): TRIPLE   0   0
              CONSTRAINT(S): PEAK   7   0
              CONSTRAINT(S): ;
              COMMAND: GENERATE
```

Figure 12   38 STRUCTURES GENERATED

Summary

In summary CASE is a highly interactive network of computer programs for reliably and efficiently assisting the chemist in the conversion of chemical and spectroscopic data to molecular structure. Communication is in the conventional language of the chemist and program execution is sufficiently rapid to make problem solving a highly conversational process. CASE is designed to grow and expand, and we are confident it will be more powerful tomorrow than it is today.

Literature Cited

1. Stevens, Calvin L., Taylor, K. Grant, Munk, Morton E., Marshall, W. S., Noll, Klaus, Shah, G. D., Shah, L. G. and Uzu, K., J. Med. Chem. (1965), $\underline{8}$, 1.
2. Munk, Morton E., Sodano, Charles S., McLean, Robert L. and Haskell, Theodore H., J. Am. Chem. Soc. (1967), $\underline{89}$, 4158.
3. Munk, Morton E., Nelson, Denny B., Antosz, Frederick J., Herald, Jr., Delbert L. and Haskell, Theodore H., J. Am. Chem. Soc. (1968), $\underline{90}$, 1087.
4. Nelson, D. B., Munk, M. E., Gash, K. B. and Herald, Jr., D. L., J. Org. Chem. (1969), $\underline{34}$, 3800.
5. Antosz, F. J., Nelson, D. B., Herald, Jr., D. L. and Munk, M. E., J. Am. Chem. Soc. (1970), $\underline{92}$, 4933.
6. Nelson, D. B. and Munk, M. E., J. Org. Chem. (1970), $\underline{35}$, 3832.
7. Nelson, D. B. and Munk, M. E., J. Org. Chem. (1971), $\underline{36}$, 3456.
8. Bognar, R., Sztaricskai, F., Munk, M. E. and Tamas, J., J. Org. Chem. (1974), $\underline{39}$, 2971.
9. Woodruff, H. B. and Munk, M. E., J. Org. Chem. (1977), $\underline{42}$, 0000.
10. Woodruff, H. B. and Munk, M. E., Anal. Chim. Acta/ Computer Techniques and Optimization, in press.
11. Shelley, C. A. and Munk, M. E., J. Chem. Inf. Comput. Sci. (1977), $\underline{17}$, 0000.
12. Ireland, C., Stallard, M. O., Faulkner, D. J., Finer, J. and Clardy, J., J. Org. Chem. (1976), $\underline{41}$, 2461.
13. Ichihara, A., Shiraishi, K., Sato, H., Sakamura, S., Nishiyama, K., Sakai, R., Furusaki, A. and Matsumoto, T., J. Amer. Chem. Soc. (1977), $\underline{99}$, 636.

8

CHEMICS: A Computer Program System for Structure Elucidation of Organic Compounds

TOHRU YAMASAKI, HIDETSUGU ABE, YOSHIHIRO KUDO,
and SHIN-ICHI SASAKI

Miyagi University of Education, Aoba, Sendai 980 Japan

There have been many articles concerned with computer programs for structure elucidation of organic compounds by analyzing chemical spectra. The methodologies and the techniques employed for this purpose can be classified into two categories, one is the identification of unknown compounds by the retrieval method of filed spectra[1,2] is carried out and the other is the generation of structural formula based on the analytical results of spectral data and other chemical evidence[3,4,5].

As reported previously, our integrated computer system for structure elucidation of organic compounds named CHEMICS stands mainly on the latter methodology [6]. IR and ^{1}H NMR spectral data of an organic compound are analyzed and plausible structural formula consistent with the analytical results are generated.

Since generation of correct structure is the major premise of this system, rather ample allowance for elucidation of partial structures is made during data analysis.

Thus, an excessive number of candidate structures (informational homologues) are generated upon occasion. In order to prevent this undesirable situation, two different strategies are considered to be practical. They are;
1) Application of the file retrieval method as a complement to the data analysis, and 2) introduction of other kinds of information sources and/or improvement of the spectral data analysis more precisely.
The former solution has been already actualized as CHEMICS-F as shown in Fig. 1 [7].

For the latter strategy, several trials have been made at our laboratory, for example, quantitative analysis of IR spectra[8], spectral simulation of ^{1}H NMR(

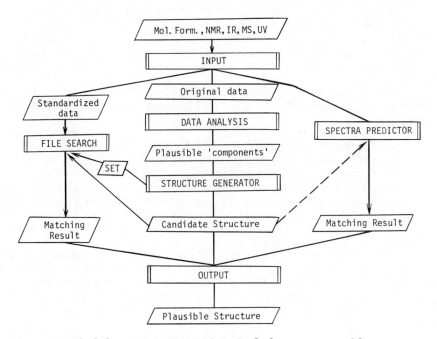

Figure 1. Block diagram of CHEMICS-F. Dashed arrow means off-line routine.

ALACON) (9), analysis of nuclear double resonance data
($^1H\{^1H\}$, NMDR) (10) and prediction of ^{13}C NMR spectra (
11).
    In this paper we describe incorporation of ^{13}C
NMR spectral analysis into CHEMICS to extend its capabilities.

General feature of ^{13}C NMR spectral data analysis

    Recently, $13C$ NMR spectroscopy has been effec-
tively employed for structure elucidation of organic
compounds.   Here we intend to introduce the spectral
data as a new information source because of its generally applicable nature.   The entire system is shown
in Fig. 2.
    The program for analysis of ^{13}C NMR spectra (
ASSINC) is composed of the following four elements as
shown in Fig. 2.
    a) DATA INPUT
    b) PRIMARY ANALYSIS
    c) SECONDARY ANALYSIS
    d) CHEMICAL SHIFT TABLE
    The idea of ASSINC is much the same as that of ^1H
NMR data analysis of the system CHEMICS (ASSIN) (6), in
which knowledge obtained by analyzing spectral data of
unknown compounds is represented as a group of substructures named 'components'.
    According to this idea, 189 kinds of 'components'
are previously defined for the ASSINC as shown (partially) in Table I, instead of the 179 'components'
for the former edition.   Each 'component' is defined
by its adjacent atoms and/or functional groups bonded
with it.

    DATA INPUT.   Input data for ^{13}C NMR data analysis
consist of positions and intensities of every signal
and their multiplicities.   We use the example of
structure 1, $C_9H_{14}O$, whose spectrum is shown in Fig. 3.

Both card and paper tape image data are
acceptable.   Even if the multiplicities
are not available, the ASSINC can analyze
the rest of the data and will offer usable
answers for successive routines.   But in
such a case, some ambiguities could not be
avoidable.

1

PRIMARY ANALYSIS.   The block diagram of the primary
analysis routine is shown in Fig. 2.   As shown in
this figure, it consists of two major parts.   One is
allocation of carbons to each spectral signal and the
other is examination of the presence of 'components'.

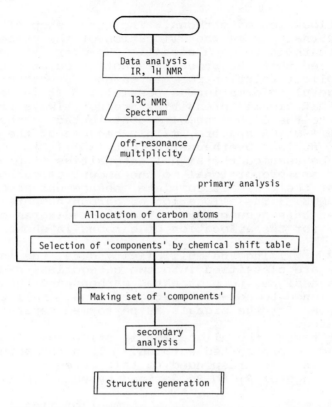

Figure 2. Flow chart of ^{13}*C NMR spectral data analysis*

NO	POSITION(ppm)	INTENSITY	MULTIPLICITY
1	24.4	1679	Q
2	28.3	4549	Q
3	33.5	895	S
4	45.2	2380	T
5	50.8	2119	T
6	125.4	2494	D
7	159.9	1084	S
8	199.2	861	S

Figure 3. ^{13}*C NMR data of compound* 1

Allocation of carbons. The first step of the primary analysis is the allocation of the proper number of carbons to each signal. However, it must be emphasized that the process is not aimed at obtaining the explicit solution for all cases, but gathering as much useful information as possible. It is well known that signal intensities are not always proportional to the carbon numbers contributed to the signals in ^{13}C NMR spectra, mainly because of the presence of nuclear Overhauser effect (NOE)(12). However, it can be assumed the signal intensities of protonated carbons are proportional to the amount of carbons because of their almost complete enhancement according to the NOE. The allocation of carbon numbers is based on this assumption. The block diagram of the routine for the allocation of carbons is shown in Fig. 4.

By utilizing the multiplicity data, the input signals are classified into two categories, namely, signals assigned to protonated carbons and those which are assigned to non-protonated carbons. Allocation of carbons for the signals is performed separately for each category.

At first, the allocation is tried for signals assigned to protonated carbons. Then the amount of carbons (AOC) corresponded to this category is limited in the range of R_1 to R_2 defined by equation (1).

$$R_1 = \left(\begin{array}{l}\text{whole carbon numbers}\\\text{of the molecule}\end{array}\right) - \left(\begin{array}{l}\text{number of signals}\\\text{assigned to non-}\\\text{protonated carbon}\end{array}\right)$$

$$\dots\dots\dots\dots (1)$$

$$R_2 = \left(\begin{array}{l}\text{number of signals assigned}\\\text{to protonated carbons}\end{array}\right)$$

After estimation of the AOC, the number of carbons for value of each signal (CNS) is evaluated by means of the equation (2) and a set of the CNS values is obtained with respect to each AOC value. However, if any one of the CNS value in the set is greater than 0.3 and less than 0.7, the set is abandoned to avoid an error.

$$CNS_i = INT_i / \sum_{j=1}^{AOS} (INT)_j * AOC \qquad \dots\dots\dots\dots\dots (2)$$

CNS_i: carbon number allocated to signal "i"
INT_i: intensity of signal "i"
AOC : amount of corresponding carbons

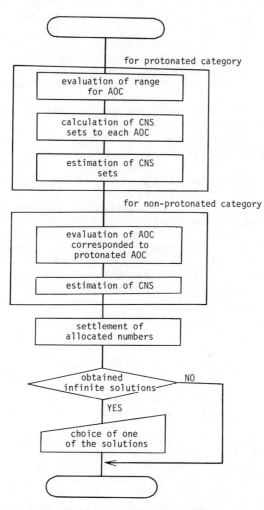

*Figure 4.   Procedure for the allocation of carbons
to each signal*

AOS : amount of corresponding signals

The allocation process for the signals assigned to non-protonated carbon is the following step.   At this stage, the AOC value is estimated in the basis of remaining carbons which are not consumed at preceeding stage.   As the result of solving the equation (2), the sets of CNS values which correspond to non-protonated carbons are obtained.   Here, it is assumed that the weakest intensity of the signal is shared with a unit number (1,2,3,...) of carbons.   Consequently, allocated numbers, namely, a set of entire CNS is acquired for each input signal.   If there is more than one solution for this problem, any one of them could be chosen as a correct set of allocated numbers to the signals.

The application of the procedure to the spectrum of compound $\underline{1}$ is described below.   The input signals shown in Fig. 3 are calssified into either protonated or non-protonated category where signals number 1,2,4, 5 and 6 are grouped into the former and 3,7 and 8 are grouped into the latter.   Through the procedure of protonated category the AOC is appraised as 5 and 6 because R_1 is calculated as 6 (9-3) and R_2 is equal to 5.   'The corresponding sets of the CNS are shown below where each integer value enclosed by parenthesis is allocated number of carbons.

signal number	1	2	4	5	6
AOC=5	0.63 (*)	1.72 (2)	0.90 (1)	0.80 (1)	0.94 (1)
AOC=6	0.76 (1)	2.06 (2)	1.08 (1)	0.96 (1)	1.13 (1)

Since it is impossible to allocate carbons to signal number 1 at the first set, this set is abandoned.   Therefore only one solution is derived from the case where the AOC is equal to 6.   At the following stage, the AOC for non-protonated category is fixed to 3, and so each residual signal must be allocated to one carbon individually.
The final result of allocated number is as follows:

signal number	1	2	3	4	5	6	7	8
allocated number	1	2	1	1	1	1	1	1

Examination of the presence of 'components'. Now we have confirmed two kinds of information about a given ^{13}C NMR spectral data. They are the amount of carbons assigned to each signal and nature of carbons (protonated or non-protonated). By considering the information, the possible presence of each 'component' is examined and those which are inconsistent with the information are abandoned.

The presence of each 'components' is judged to be appropriate by its chemical shift range (refer to Table I), in other words, if there are no signals within a chemical shift range corresponding to a 'component', it is judged to be not present in a sample compound.

As shown in Fig. 5, twenty-nine components survive for compound 1, through the primary analysis. The result of the primary analysis is represented by the matrix named NM matrix, in which each row is corresponding to a survived 'component' and each column to each signal of the given ^{13}C NMR spectrum. Each matrix element indicates maximum number of the carbons for 'component' assigned to the corresponding signal. Those elements with value -1 indicate corresponding 'components' were not assigned to the signals.

SECONDARY ANALYSIS. At the first step of this routine, a set of 'components' which is consistent with the molecular formula is selected from survived 'components'. One of the five sets which was finally generated for compound 1 is shown in Fig. 6.

As described before, each of the signals is treated as if it were independent of the others and the 'components' which can be assigned to at least one signal survive without any further examination at the primary analysis.

However, it is necessary to examine whether the set is consistent with the given spectrum or not, in other words, each of all 'components' of the set should be confirmed whether they are fully consistent with the input spectrum with neither excess nor deficiency.

To make this examination, the selective NM matrix is made for the set by extracting the rows corresponding to selected 'components' from NM mateix shown in Fig. 5. This selective matrix is shown in Fig. 6.

As shown in Fig. 7, this matrix N is converted into another matrix X by substituting the positive elements by variables (x_{ij}) and the negative elements by zeros. A set of simultaneous linear equations is made from X and two constnat vectors C and D, repre-

NO	CMP	SUB/STRUCTURE			NM MATRIX						
1	10	GEM-DIMETHYL-(D)			1	2	-1	-1	-1	-1	-1	-1
2	11	GEM-DIMETHYL-(T)			1	2	-1	-1	-1	-1	-1	-1
3	12	GEM-DIMETHYL-(C)			1	2	-1	-1	-1	-1	-1	-1
4	14	CH3-(C)-	(Y)		1	2	-1	-1	-1	-1	-1	-1
5	17	CH3-(C)-	(T)		1	2	-1	-1	-1	-1	-1	-1
6	33	CH3-	(D)		1	2	-1	-1	-1	-1	-1	-1
7	38	CH3CO-	(D)		1	-1	-1	-1	-1	-1	-1	-1
8	40	CH3CO-	(C)		1	2	-1	-1	-1	-1	-1	-1
9	106	-CH2-	(C)	(K)	-1	-1	-1	1	1	-1	-1	-1
10	107	-CH2-	(C)	(D)	-1	-1	-1	1	1	-1	-1	-1
11	108	-CH2-	(C)	(T)	-1	-1	-1	1	1	-1	-1	-1
12	109	-CH2-	(C)	(C)	-1	-1	-1	1	1	-1	-1	-1
13	118	-CH=<OLEFIN>			-1	-1	-1	-1	-1	1	-1	-1
14	143	:C= <OLEFIN>			-1	-1	-1	-1	-1	-1	1	-1
15	144	=C= <KETENE>			-1	-1	-1	-1	-1	-1	-1	1
16	145	=C= <ALENE>			-1	-1	-1	-1	-1	-1	-1	1
17	146	FURAN(O)										
18	153	-O-										
19	172	-CO-	(C)	(D)	-1	-1	-1	-1	-1	-1	-1	1
20	173	-CO-	(C)	(T)	-1	-1	-1	-1	-1	-1	-1	1
21	174	-CO-	(C)	(C)	-1	-1	-1	-1	-1	-1	-1	1
22	175	O=C=			-1	-1	-1	-1	-1	-1	-1	1
23	177	Y	(O)		-1	-1	-1	-1	-1	-1	1	-1
24	182	Y	(C)		-1	-1	-1	-1	-1	-1	1	-1
25	184	C	(Y)		-1	-1	1	-1	-1	-1	-1	-1
26	185	C	(K)		-1	-1	1	-1	-1	-1	-1	-1
27	186	C	(D)		-1	-1	1	-1	-1	-1	-1	-1
28	187	C	(T)		-1	-1	1	-1	-1	-1	-1	-1
29	188	C	(C)		-1	-1	1	-1	-1	-1	-1	-1

SAMPLE =   1

JOB END

Figure 5. Survived components of compound 1 through ^{13}C NMR data analysis

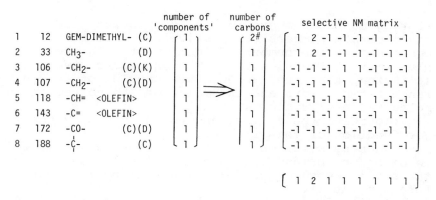

number of number of
'components' carbons selective NM matrix

allocation number of
carbons

\# Only methyl carbon is considered in gem-dimethyl group

Figure 6. Selective components for the fifth set of compound 1

N:
$$\begin{pmatrix} 1 & 2 & -1 & -1 & -1 & -1 & -1 & -1 \\ 1 & 2 & -1 & -1 & -1 & -1 & -1 & -1 \\ -1 & -1 & -1 & 1 & 1 & -1 & -1 & -1 \\ -1 & -1 & -1 & 1 & 1 & -1 & -1 & -1 \\ -1 & -1 & -1 & -1 & -1 & 1 & -1 & -1 \\ -1 & -1 & -1 & -1 & -1 & -1 & 1 & -1 \\ -1 & -1 & -1 & -1 & -1 & -1 & -1 & 1 \\ -1 & -1 & 1 & -1 & -1 & -1 & -1 & -1 \end{pmatrix}$$

X:
$$\begin{pmatrix} x_{11} & x_{12} & 0 & 0 & 0 & 0 & 0 & 0 \\ x_{21} & x_{22} & 0 & 0 & 0 & 0 & 0 & 0 \\ 0 & 0 & 0 & x_{34} & x_{35} & 0 & 0 & 0 \\ 0 & 0 & 0 & x_{44} & x_{45} & 0 & 0 & 0 \\ 0 & 0 & 0 & 0 & 0 & x_{56} & 0 & 0 \\ 0 & 0 & 0 & 0 & 0 & 0 & x_{67} & 0 \\ 0 & 0 & 0 & 0 & 0 & 0 & 0 & x_{78} \\ 0 & 0 & x_{83} & 0 & 0 & 0 & 0 & 0 \end{pmatrix}$$

c^T: (2 1 1 1 1 1 1 1)   D: (1 2 1 1 1 1 1 1)

a) N, X, C, and D mean selective NM matrix, selective NM matrix replaced by x_{ij}, modified 'component' vector and allocation vector, respectively.

$$\begin{cases} X \cdot I = C \\ I \cdot X = D \end{cases}$$

b) representation of simultaneous linear equations where I means unit row vector having eight elements.

Figure 7. Representation of simultaneous linear equations for the fifth set of compound 1

senting carbon numbers in the 'components' and allocated carbon numbers, respectively.   The number of
equations is the number of 'components' in the set
plus that of the signals.   The equations have a re-
striction, that the variable x_{ij} should not exceed the
range between zero and the value of the corresponding
selective matrix element.

To solve these simultaneous equations is the major function of this routine.

When no solution is obtained, the set is judged
to be inappropriate one, and when a solution is given,
the set is sent to the following routine (the structure generator).

At the final stage of the spectral analysis, five
sets of components which are generated from twenty-
nine components are selected as plausible ones for
compound 1.   Five sets are shown as follows, numer-
al in parenthesis expresses number of the component;

NO. 1   10(1), 38(1), 107(1), 109(1), 118(1), 143(1), 189(1)

NO. 2   10(1), 40(1), 106(1), 107(1), 118(1), 143(1), 189(1)

NO. 3   12(1), 38(1), 107(2), 118(1), 143(1), 189(1),

NO. 4   10(1), 33(1), 106(1), 109(1), 118(1), 143(1), 172(1),
         189(1)

NO. 5   12(1), 33(1), 106(1), 107(1), 118(1), 143(1), 172(1),
         189(1)

The overall process that 189 components are reduced into 29 by means of the examination of molecular
formula followed by the successive analyses of IR, [1]H
NMR and ^{13}C NMR is shown in Fig. 8.   In fig. 8, nu-
merals 108, 105, 59 and 29 in parentheses indicate the
amounts of survived components by successive restrictions of molecular formula, IR, ^{1}H NMR and ^{13}C NMR,
respectively.   Only five sets of components uncontra-
dictory with molecular formula and given NMR spectrum
are picked up from these twelve components.   Finally,
the structure generator(13) is applied to generate the
structures from each set of components so that 3, 1,
2, 3 and 3 structure(s) produced for sets, 1, 2, 3, 4
and 5.   These structures are shown in Fig. 9 as informational homologues for the input molecular formula
and chemical spectra.
The underlined one is the structure of the compound 1.

PREPARATION OF CHEMICAL SHIFT TABLE. A chemical
shift ranges for a signal of a 'component' was determined for the analysis described in the previous

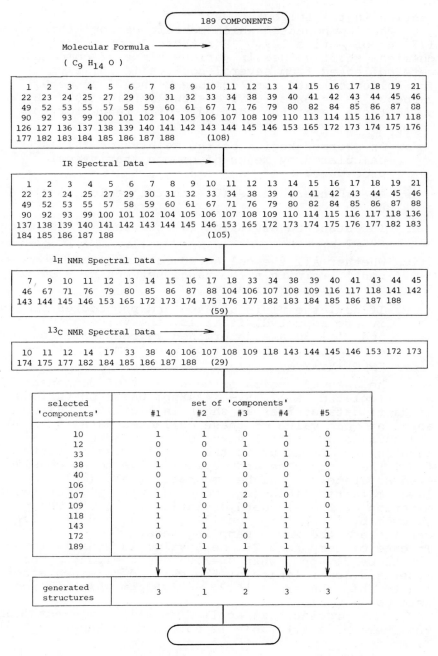

Figure 8. Feature of reducing the number of components through consecutive analyses of compound 1

section in the following way.

The 'components' which contain carbon atoms are 177 out of entire 189. For those 'components', their chemical shift values in various kinds of compounds were collected from several sources(14,15,16). The collected data for 'component' no.25 of methyl carbons, as an example, are shown in Fig. 10.   By using these data, the chemical shift range for the 'component' is obtained as follows.

i. An assumed region of the mean value ($\bar{\mu}$) is calculated by means of common statistical procedure.

ii. An arbitrary value ($\bar{\mu}'$) in the region is picked up.

iii. The standard deviation (σ) for the $\bar{\mu}'$ is calculated.

iv. Whether all the collected data for the 'component' are within the range between $\bar{\mu}' - 3\sigma$ to $\bar{\mu}' + 3\sigma$ is examined.

v. If not, the $\bar{\mu}'$ is updated and procedures iii and iv are repeated, if it is, the values $\bar{\mu}' - 3\sigma$ and $\bar{\mu}' + 3\sigma$ are determined as the upper and lower limits of the shift of the 'component' respectively.

The assumed region of mean value of component 25 was calculated as 19.15 - 24.48ppm based on various kinds of data sources as shown in Fig. 10.   Here, an apparent mean value of these collected data is 21.8ppm and this is an initial value of $\bar{\mu}'$.   Some data of samples are often out of the normal Gaussian distribution, therefore standard deviation has to be considered separately in higher magnetic field (σ_H) and lower magnetic field (σ_L) compared with $\bar{\mu}'$, for determination of the standard deviation for $\bar{\mu}'$.   The $\bar{\mu}'$ is renewed by 'flip-flop' untill $\bar{\mu}' - 3\sigma_H$ and $\bar{\mu}' + 3\sigma_L$ can include the whole sampling data.
In case of component 25, mean value is finally found out to be 21.4ppm, when $\sigma_H = 2.05$ and $\sigma_L = 1.39$.   The upper and lower limits of the shift determined according to this manner is 15.21 - 25.53ppm which is registered in Table I.

This procedure is applied to all 'components' and the chemical shift table is obtained as shown in Table I.

Result and Discussion
The result obtained for twenty two compounds by

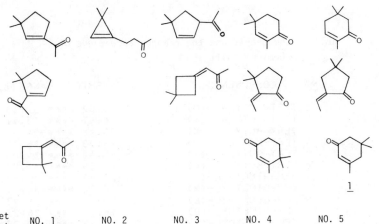

from the set
of components   NO. 1   NO. 2   NO. 3   NO. 4   NO. 5

Figure 9. Structures generated from each set of compound 1

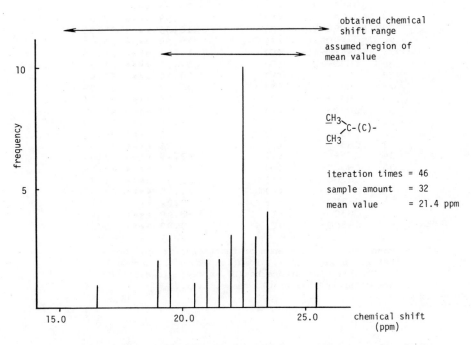

Figure 10. Estimation of ¹³C NMR chemical shift range of component #25

Table I   Components and their appearance range of ^{13}C NMR
          chemical shift

NO	COMPONENT		SHIFT RANGE (ppm)
1	TERT-BUTYL-	(O)[#]	26.02 **** 31.13
2	TERT-BUTYL-	(Y)	24.47 **** 33.57
3	TERT-BUTYL-	(K)	25.48 **** 34.04
4	TERT-BUTYL-	(D)	28.23 **** 36.78
5	TERT-BUTYL-	(T)	25.48 **** 34.04
6	TERT-BUTYL-	(C)	23.65 **** 32.97
7	GEM-DIMETHYL-	(O)	27.42 **** 32.95
8	GEM-DIMETHYL-	(Y)	10.72 **** 36.27
9	GEM-DIMETHYL-	(K)	10.12 **** 36.27
10	GEM-DIMETHYL-	(D)	14.72 **** 36.27
11	GEM-DIMETHYL-	(T)	10.12 **** 36.27
12	GEM-DIMETHYL-	(C)	6.80 **** 32.61
13	CH3-(C)-	(O)	4.58 **** 32.01
14	CH3-(C)-	(Y)	4.58 **** 32.01
15	CH3-(C)-	(K)	5.25 **** 15.50
16	CH3-(C)-	(D)	10.43 **** 21.53
17	CH3-(C)-	(T)	4.58 **** 32.01
18	CH3-(C)-	(C)	9.92 **** 12.97
19	ISO-PROPYL-	(O)	15.09 **** 25.83
20	ISO-PROPYL-	(A)	16.63 **** 25.83
21	ISO-PROPYL-	(Y)	20.95 **** 25.45
22	ISO-PROPYL-	(K)	15.09 **** 23.87
23	ISO-PROPYL-	(D)	16.33 **** 25.83
24	ISO-PROPYL-	(T)	15.09 **** 25.83
25	ISO-PROPYL-	(C)	15.21 **** 25.53
26	CH3O-	(O)	52.88 **** 61.61
27	CH3O-	(Y)	54.59 **** 57.92
28	CH3O-	(K)	50.34 **** 52.53
29	CH3O-	(D)	56.68 **** 61.51
30	CH3O-	(T)	52.88 **** 61.51
31	CH3O-	(C)	49.95 **** 60.60
32	CH3-	(Y)	7.26 **** 26.10
33	CH3-	(D)	7.06 **** 33.08
34	CH3-	(T)	-2.49 **** 8.49
35	CH3CO-	(O)	19.81 **** 23.39
36	CH3CO-	(Y)	22.95 **** 31.79
37	CH3CO-	(K)	22.95 **** 33.92
38	CH3CO-	(D)	22.22 **** 28.15
39	CH3CO-	(T)	-2.49 **** 8.49
40	CH3CO-	(C)	20.80 **** 30.01

means the adjacent atom or functional group, they are,
  saturated oxygen (O), aromatic carbon(Y), carbonyl
  carbon(K), olefinic carbon(D), acetylenic carbon(T),
  and saturated carbon(C), respectively.

Table II   Results obtained for several compounds by CHEMICS

no.	compound	molecular formula C	H	O	number of IH through IR, ^{1}H NMR analysis	number of IH through IR, ^{1}HNMR,^{13}CNMR analysis
1	α-Methyltetrahydrofuran	5	10	1	10	1
2	p-Quinone	6	4	2	589	2
3	2-Methylpentane	6	14		3	1
4	3-Methylpentane	6	14		3	1
5	2,3-Dimethylbutane	6	14		4	1
6	3-Heptanone	7	14	1	3	2
7	2-Heptanone	7	14	1	4	1
8	m-Xylene	8	10		40	21
9	Ethylbenzene	8	10		5	5
10	Cyclohexylacetate	8	14	2	161	1
11	2-Octanol	8	18	1	38	1
12	Coumarine	9	6	2	834	116
13	Isophorone	9	14	1	1895	12
14	Diisobutylketone	9	18	1	30	1
15	n-Nonanol	9	20	1	24	1
16	Dicyclopentadiene	10	12		1729	41
17	Verbenone	10	14	1	53274<	42
18	Camphor	10	16	1	3253	75
19	n-Decanol	10	22	1	50	1
20	2-Cyclohexylcyclohexanone	12	20	1	2109	147
21	β-Ionone	13	20	1	57827<	481
22	Methylmyristate	15	30	2	4767	1

Table III   Results obtained by utilizing various combinations of information
sources

compound	C	H	O	analytical mode*	number of 'informational homologues'
3-Heptanone	7	14	1	P C P+C C+O P+C+O	3 6 3 2 2
2-Octanol	8	18	1	P C P+C C+O P+C+O	38 41 13 1 1
Isophorone	9	14	1	P C P+C C+O P+C+O	1895 27 12 27 12

* P, C, and O mean an analysis of ^{1}H NMR, ^{13}C NMR and/or
off-resonance spectra, respectively.

means of the old system and new system are presented in Table II.   The correct structure has been always generated among the plausible structures.   The numbers of informational homologues obtained by means of ASSINC are reduced to 19.7 percent (in simple average) or 3.1 percent (in weighted average) of those obtained by means of the old system where only IR and ^{1}H NMR data were analyzed.   As a result of the addition of ^{13}C NMR spectral data analysis, it becomes possible to decrease remarkably the numbers of informational homologues. For example, the number was reduced from 4767 to one for compound 20 as shown in Table II.

Table III shows the number of informational homologues of several compounds obtained by utilizing various combinations of information sources, namely, ^{1}H NMR, ^{13}C NMR, ^{1}H NMR plus ^{13}C NMR, ^{13}C NMR plus its off-resonance data, and ^{1}H NMR plus ^{13}C NMR plus off-resonance data.   As shown in this table, the number of informational homologues and the number of the component sets are both decreased in accordance with the addition of new information sources.

In conclusion, the number of the 'informational homologues' and the 'component' sets are satisfactorily reduced by consecutive analyses.   As mentioned above, the efforts to reduce the excessive 'components' bear good fruits, i.e., the number of the produced sets are less than ten for all cases.   Therefore, the information about the conectivities between all the 'components' in a set become important data to include in a future system.
That kind of information will work effectively to reduce the excessive 'informational homologues' and nuclear magnetic resonance techniques will give such information.

This work was supported in part by a Scientific Research Grant from the Ministry of Education, Japan.

Literature Cited

(1)   Schwarzenbach,R.,Meili,J.,Koenitzer,H. and Clerc, J.T., Org. Mag. Resonance, (1976),8,11
(2)   Bremser,W.,Klier,M. and Meyer,E., ibid, (1975), 7,97
(3)   Carhart,R.E.,Smith,D.H.,Brown,H. and Djerassi,C., J. Am. Chem. Soc., (1975), 97, 5755
(4)   Beech,G.,Jones,R.T. and Miller,K., Anal. Chem., (1976), 46, 714
(5)   Gray,N.A.B., ibid, (1975), 47, 2426
(6)   Sasaki,S. et al, Mikrochimica Acta(Wien), (1971), 726

(7) Sasaki,S., CHEMICS-F in "Information Chemistry", p227, The University of Tokyo Press, Tokyo, 1975, and the detail of CHEMICS-F will be reported in the near future.
(8) Miyashita,Y. and Sasaki,S., Jpn.Chem.Soc. Meeting, (1975), I, 174
(9) Yamasaki,T. and Sasaki,S., Jpn. Anal., (1975),213
(10) unpublished
(11) Ochiai,S.,Hirota,Y.,Kudo,Y. and Sasaki,S., Jpn. Anal., (1973), 22, 399
(12) Stother,J.B., "Carbon-13 NMR Spectroscopy", Academic Press, New York, 1972
(13) Kudo,Y. and Sasaki,S., J.Chem. Inf. Comput. Sci., (1976), 16, 43
(14) Beach,L.B., "API 44 Selected ^{13}CNMR Spectral Data" API Research Project 44 Publication, Texas, 1975
(15) Nuclear Magnetic Resonance Spectral Search System, NIH/EPA, USA
(16) Johnson,L.F. and Jankowski,W.C., "Carbon-13 NMR Spectra", Wiley Interscience, New York, 1972

9

Computer Assistance for the Structural Chemist

RAYMOND E. CARHART—Department of Computer Science,
Stanford University, Stanford, CA 94305

TOMAS H. VARKONY—Department of Chemistry,
Stanford University, Stanford, CA 94305

DENNIS H. SMITH—Department of Genetics,
Stanford University, Stanford, CA 94305

Elucidation of unknown molecular structures can be thought of as a process of systematic posing, testing and rejection of hypotheses. Each hypothesis is of course a partial or complete structure which is evaluated in light of available evidence. Chemists perform these tasks quite well if not completely systematically. We seek computer programs which emulate processes of reasoning about chemical structure to save time, to stimulate the chemist's thinking about an unknown and to guarantee that no plausible alternatives have been overlooked. There are several areas of structure elucidation which are amenable to some degree of computer assistance. Computer techniques are employed routinely to aid in collection and preliminary analysis of data from several types of spectrometers. Applications of problem solving programs to more sophisticated analysis of molecular structure are more recent. For example, several reports have appeared describing computer programs for assisting chemists in the task of constructing plausible candidate structures for unknown compounds (1-3). These programs have matured to the point where successful applications to real-world structural problems have been demonstrated (1,4). The structure generating capabilities of these programs fulfill only the generate phase of the "plan-generate-test" paradigm of heuristic search (5). Much work remains to be done on the structure generators to make them simpler to use and to reduce them in size, complexity and execution time. However, this work is primarily developmental. In this report we focus on new research on "planning" and "testing" ("predicting"), and illustrate how computer techniques can provide assistance in these areas also.

We define planning to include initial interpretation of chemical and spectroscopic data and the translation of results into fragments of molecular structure which can be used manually, or together with a structure generator, to generate structures (2, 6-12 and other papers in this symposium). Computer assistance can be provided to the chemist during

126

planning in several ways. We discuss in the subsequent section ways in which our structure generator, CONGEN, is being modified to plan more efficient structure generation by translation of structural data input to the program. We feel that such intelligent use of structural data as constraints is essential during structure generation.

In most problems, however, initial data and preliminary analysis are insufficient to determine a structure uniquely; many, perhaps hundreds, of candidate structures may remain. Therefore, we are also pursuing development of computer methods to examine or evaluate large sets of candidate structures to assist the chemist in focussing on the correct structure. These methods involve: a) testing of structures with constraints; and b) prediction of results of manipulation of the structures which can be matched against actual laboratory measurements. The goal in both methods is to reject implausible candidate structures. We describe recent results in a subsequent section. We feel such computer-aided facilities will be useful in discriminating among a large set of candidate structures.

I) Planning via Constraint Interpretation

We have previously described the CONGEN program for constructing structures under constraints (3). The program has an extensive repertoire of constraints which prevent the generating algorithm from producing undesired partial or complete structures (13). It has a rich language for defining partial structures (substructures) including several terms describing atom and bond properties, rings, chains, and so forth. The language allows representation of aromaticity, atoms of indeterminate, or restricted, identity and chains and rings of variable numbers of atoms. CONGEN is an interactive program with a helpful user interface with extensive error checking to avoid program crashes. It is useful, and used by a nationwide community of persons via a computer network, even though it only fulfills the role of a structure generator. It depends on user inputs to determine its knowledge of chemistry and, until recently, utilized this information much as it was received. This is not the most efficient way to solve problems. Much needs to be known about the program to use it in the most efficient way for a given problem. In modifications and extensions to CONGEN (below) we are striving to divorce the chemist from the algorithms. We seek to solve problems efficiently, beginning with information supplied by the chemist to the program in ways which are intuitive to him/her, independent of the program.

Other workers have discussed aspects of data interpretation at some length (2, 6-12). We are directing our attention to the problem of how to translate and utilize the structural information provided by these interpretations. All structural information about an unknown can be viewed as constraints. Such constraints may be positive statements involving substructures or ring systems which must be present, or negative statements forbidding the presence of other substructures or ring systems. There is sometimes a significant conceptual gap between the intuitive chemical phrasing of a CONGEN problem and the phrasing for most efficient problem solving by the program. There are usually many ways of defining a given problem and different definitions can place widely different demands upon the program. We have a continuing interest in reducing this conceptual gap by making CONGEN responsible for rephrasing a problem in an efficient way, thus freeing the chemist to concentrate upon the chemical, rather than the algorithmic, aspects of a given problem. We have recently described some initial efforts toward automatic rephrasing of problems, e.g., implications of ranges of hydrogen atoms on a given atom (13). We have so far treated in detail only the problem of translation of GOODLIST (13) constraints (required structural features). Although this is only a part of the constraints interpretation problem, the resulting algorithm is a significant step forward in development of our structure generating capabilities for reasons discussed below.

A. GOODLIST Interpretation. Constructive Substructure Search. All current structure generation programs, including until now CONGEN, have a serious limitation. Substructures used as building blocks, or "superatoms" (3) are required to be non-overlapping i.e., they must have no atoms in common. In many structural problems several large fragments of the structure may be known, but the overlaps among them are not known. The chemist (or the program (2)) must decide what set of fragments cannot overlap and use those as superatoms. Larger fragments which may overlap are tested at the end of structure generation by a graph-matching procedure. This distinction is conceptually inelegant and puzzling to chemists who wish to use CONGEN. Lack of understanding of the distinction can lead to terrible inefficiencies. Inefficiency arises in two ways. A set of small superatoms will yield many more final structures than a set of large superatoms. A great many of the structures produced by the former set will be rejected on post-testing (graph-matching) with the larger GOODLIST fragments. These computations require considerable time. But chemists must contend with the problem of overlapping substructures in their manual exploration of structural possibilities. By examining how such problems are solved manually we have developed a new procedure which employs both the efficiency of non-overlapping superatoms and the efficiency of incorporating GOODLIST information at the beginning of the process of structure generation (thus, "planning") rather than the end. In short, we have developed a general solution to the problem of constructing structures from fragments which overlap to an unspecified degree.

   1) An Example. A recent problem studied using CONGEN involved the structural information summarized in Figure 1. The bonds with unspecified termini in CEMB are free valences. All atoms (excepting seven remaining hydrogens) in the empirical formula $C_{20}H_{24}O_1$ are included in CEMB. The problem consists of finding all ways of allocating three new bonds among the free valences in the superatom CEMB such that the three indicated substructures (Figure 1) are present in the final molecules. There are perhaps 10,000 unique allocations of those three new bonds, but only seven pass the GOODLIST tests. Using GOODLIST as a post-test only, CONGEN would generate all 10,000 and discard nearly all of them, a process which would have been so lengthy that it was never completed.

   Yet most chemists given this information could fairly quickly write down the seven solutions. It is clear conceptually how the problem is solved manually. It is obvious that there are only three places in CEMB where the first GOODLIST item (Figure 1) can fit, because there are only three methyl groups on the periphery of CEMB which fit the structural criteria of the substructure. For each of these matchings, there are four ways of matching the second substructure, and so forth. In this case, some matchings lead to construction of new bonds. In the general case when there are several superatoms and remaining atoms, matchings may require constructing new bonds and structural units, by utilizing parts of existing superatoms or remaining atoms.

   In this example, CONGEN, using the method outlined below, solves the problem in much the same way as described above for the manual method. Rather than generating and testing thousands of undesired structures, it quickly arrives at the seven solutions by incorporating the GOODLIST information from the very beginning.

   2) The Method. We have developed a method, as an extension to CONGEN, which emulates the manual method. It matches GOODLIST items to the initial problem formulation, which may be anything from the raw empirical formula to a list of non-overlapping superatoms. When new bonds or atoms are required to complete the matching, they are constructed from other segments (atoms, superatoms) of the problem by forming new bonds. In order to incorporate a GOODLIST substructure it is necessary to find all unique ways that the given substructure can be created using parts of the existing building blocks (atoms and superatoms). Figure 2 shows schematically, together with an example, some of the ways this construction might occur: a) by bonding together two (or more) existing superatoms to create one larger one; b) by bonding additional atoms to a superatom to

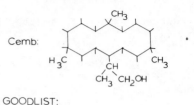

GOODLIST:

Figure 1. Structural information available for an unknown cembranolide. The superatom "CEMB" was inferred from a variety of data and inferences derived from related, co-occurring compounds. The GOODLIST items were inferred from additional spectroscopic and chemical data.

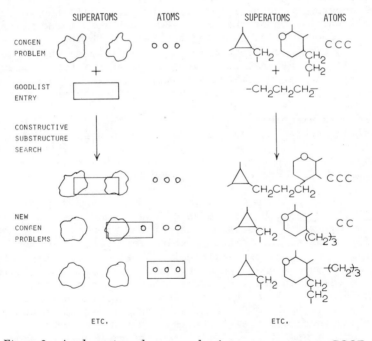

Figure 2. A schematic and an example of ways to construct a GOOD-LIST item from a collection of atoms and superatoms. Each way results in larger superatoms in new problems to be carried on to the next step in structure generation, for example, incorporation of another GOOD-LIST item.

create a larger one; and c) by constructing a copy of the substructure from single atoms, creating a new superatom. We call this method "constructive substructure search" for obvious reasons.

The procedure is stepwise. It begins with finding all ways to construct the first GOODLIST item. Each of these ways is a new problem with the same or larger superatoms (e.g., Figure 2). Each new problem is treated in a depth first generation scheme by considering the next GOODLIST item, and proceeding until all GOODLIST items are accounted for or until the next GOODLIST item cannot be built from the current set of superatoms and atoms. Depending on the problem, the final set of solutions may be final structures, as in the previous example (Figure 1), or may be several new sets of superatoms and atoms for each of which CONGEN can be used to construct final structures. In the latter instance, all GOODLIST items are guaranteed to be present so that time-consuming post-testing for their existence is not required. Details of the algorithm are beyond the scope of this presentation and will be discussed separately. Briefly, the algorithm is derived from the CONGEN graph-matching routine (13) with the additional feature that as it searches for the substructure it is allowed to create new bonds (up to the limit of available new bonds in the original CONGEN problem) whenever they are necessary for the search to proceed. New bonds may be used to form multiple bonds or rings within a superatom. Alternatively, use of new bonds to form new connection to atoms outside the superatom leads to extended superatoms (Figure 2). During the search, full account is taken of the topological symmetry of the superatoms in the original problem so that fittings which are redundant with respect to these symmetries are avoided. The potential equivalence of GOODLIST items is not currently considered; duplicate problems or final structures are removed in subsequent steps. Most CONGEN problems contain one or more GOODLIST items which can be treated with our method. The algorithm is currently being tested to understand its scope and limitations. Although integrated into CONGEN, it does not yet have an appropriate user interface so is not yet available as part of our version available over the network.

3) A Second Example. The structural information for a second example is summarized in Figure 3. This information is formulated from data presented in a previous discussion of this structure (14). In this example, the initial set of non-overlapping structural components consisted of the off-resonance decoupled 13C NMR data plus the atom Z (Figure 1). Every bond connecting atoms in the final structure must be used to connect these components. To generate structures from this information alone, with subsequent testing for the presence of S1- S4 (Figure 3) would be effectively impossible. The GOODLIST items S1 -S4 can overlap significantly. The problem was also solved

by reducing the substructures S1-S4 to non-overlapping components and generating structures under extensive constraints. But this structural information can be used directly and efficiently in our new method, and CONGEN, using constructive substructure search, arrives quickly at representations of the two possible solutions, 1 and 2, Figure 4. If one assumes that the two amide functionalities S2 must be completely disjoint, the structure 2, proposed previously (14) is the only solution, excepting structural variation in Z (see caption to Figure 3).

II. Testing via PRUNE, SURVEY, REACT and MSPRED

From an algorithmic standpoint, CONGEN is successful if it can, in a reasonable amount of time and without exhausting storage resources, produce a list of candidate structures satisfying the chemist's constraints. However, this list is often quite large, perhaps several hundred structures, and from an analytical standpoint the problem may be far from complete. It remains for the chemist to discriminate among the candidates, eventually reducing the possibilities to just one structure. We are studying several ways to provide computer assistance in examining and further constraining lists of structural candidates. We have implemented several types of tests designed to help discriminate among candidate structures.

A. Direct Tests. Two tests are directly related to structural features of the candidates.

1. PRUNE. The chemist can further reduce a set of structures by "pruning" (3) the set with new structural information. This subsequent testing has been described in detail previously (3,13).

2. SURVEY. The second type of direct test allows surveying the set of candidate structures using a library of predefined structural features. The functions of SURVEY and examples of applications are summarized in Figure 5. The examples are chemical areas where we have utilized SURVEY in our own research.
An extensive library of functional groups is maintained in order that sets of structures may be surveyed for unlikely functional groupings. Such groupings are often overlooked in the initial definition of a problem precisely because they are unlikely. But CONGEN will construct them unless constraints forbid it. SURVEY acts as a reminder of such groupings. It provides a classification of structures by functional groups, any set of which can be kept or pruned away at the chemist's discretion. Other libraries of structural features are maintained, including, for example, libraries of mono- and

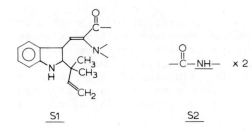

S1

$-\overset{\overset{\displaystyle O}{\|}}{C}-NH-$ × 2

S2

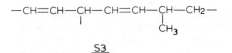

S3

$C\!\!=\!\!C-Z$

S4

Figure 3. Structure information for a dehydrotryptophan derivative. The atom Z represents an aromatic ring with several other substituents. The variety of possible substitution patterns on the aromatic ring lends considerable structural variety to this problem beyond our brief presentation. Emperical formula: $C_{26}H_{28}N_3$-O_2Z. C13 NMR: $CH_3 \times 3$, $CH_2 \times 2$, $CH \times 12$, $C \times 9$.

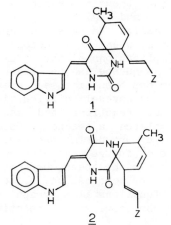

1

2

Figure 4. Two structural possibilities for the compound whose structural information is summarized in Figure 3

FUNCTION: AIDS IN PERCEPTION OF ANY OF A
PRE-SPECIFIED SET OF STRUCTURAL
FEATURES IN A GROUP OF
STRUCTURAL CANDIDATES.

E.G. A) FUNCTIONAL GROUPS
      B) TERPENOID SKELETONS
      C) AMINO ACID SKELETONS

Figure 5. Function and examples of application of SURVEY, a subprogram of CONGEN, used to test candidate structures

sesquiterpane skeletons.   We have utilized these libraries in
studies of terpenoid cyclization and rearrangement (15) in order
to locate known skeletons and elucidate their pathways of
formation. A library of amino acid skeletons is used to test
structural candidates which may be conjugates of organic acids
with amino acids in studies of organic constituents of human
body fluids. Any set of features or structures can be defined
and used by SURVEY.

     B. Indirect Tests. There are two informative sources of
data about structural candidates which cannot always be phrased
as direct tests on the candidates themselves: 1) structural
features observed in chemical reactions; and 2) empirical
spectroscopic measurements on the unknown which cannot be
interpreted unambiguously in precise structural terms. The
program REACT addresses the first problem while MSPRED is
concerned with the second in the context of mass spectrometric
observations.

     1) REACT. The REACT program has two basic goals: 1) to
provide the chemist with a computer-based language for defining
graph transformations and applying them to structures, thus
simulating chemical reactions; and 2) to keep track
automatically of the interrelationships among structures in a
complex sequence of reactions so that whenever structural claims
are made about any product, the implications of these claims on
structures at other steps in the sequence can be traced. The
first version of the REACT program has been discussed previously
(16).   Based on our experience and several difficulties with
representation of the network of reactions and associated
structures, we have recently completed a second version with
several new features. The goals we set during the writing of
this version and for near-term future developments are
summarized in Figure 6. The program has been separated from
CONGEN for purposes of efficiency (the combination is too large
a program) and because certain functions like graph matching
have a slightly different meaning when applied to reactions vs.
structure generation.
     EDITREACT, the reaction-editing language, has been extended
to allow the user to define subgraph constraints which apply
relative to a potential reaction site rather than to the
molecule as a whole. For example, in the present version of
REACT, we can say either that a hydroxyl group (OH), if present
anywhere in the reactant molecule, would inhibit the reaction,
or that such inhibition would take place only if the OH group is
adjacent to the reaction site. Such site-specific constraints,
applied either before or after the transformation (i.e.,
reaction) has been carried out on the site, are critical to the
detailed description of real chemical reactions. The inclusion
of this facility in REACT substantially increases its usefulness

in real-world chemical problems. The control structure for REACT (Figure 6, 3) has undergone major revision. In the initial implementation, a set of products arising from the application of a given reaction to a given starting structure could be subjected to a multi-level classification which grouped the products based upon user-defined substructural constraints. Each of these classes had an associated minimum and maximum number, representing the numbers of products which were allowed to be members of the class. Any starting materials whose products could not satisfy these conditions were removed from the list of candidates. Structures in any class could be further reacted, their products classified, and so on. This treatment of bookkeeping was sufficient for stating many chemical problems. For example, suppose a chemist knew that a particular reaction on an unknown compound yielded two carbonyl compounds (i.e., containing C=O), at least one of which was an ester (R-O-C=O(R´)). He could define a product class CARBONYL using the C=O substructure with a minimum and maximum of two products. He could then define a sub-class of CARBONYL called ESTERS using the substructure R-O-C=O(R´) with a minimum of one and a maximum of two products. The program would automatically use this information to eliminate candidate starting structures which could not give the indicated product distribution with the given reaction. There are chemical problems, though, for which the above scheme is too rigid. For example, suppose a reaction gives several products two of which are isolated and labeled δA and δB. Suppose that only a small amount of δA is available so only mass spectroscopic measurements are practical and that a deuterium-exchange experiment shows that δA has two exchangeable protons (say, either N-H or O-H). Presume that δB shows a strong carbonyl absorption in the IR. Now, δA might also contain a carbonyl group, but that was never determined, and neither was the number of exchangeable protons in δB, which could be two. No matter how one attempts to use the above-described classification system, one cannot express this information accurately.

Our new approach is designed to express chemical information to REACT in a much more natural sequence which parallels the experimental steps. Current operations, which include functional commands and descriptive terms, are summarized in Figure 7. Thus, a reaction is carried out by using the command REACT and applying a named reaction to a set of structures (initially STRUCS (Figure 7), but subsequently to the contents of any named product flask). Products are assigned to a named flask. The first experimental step after a reaction is usually the separation and purification of products. An analogous step is included in REACT, in which the separation amounts to defining a number of labelled "flasks" each of which is ultimately to contain a specified number (usually 1) of the products. As experimental data are gathered on each real

REACTION CHEMISTRY DEVELOPMENTS

1. SEPARATION FROM CONGEN - COMMUNICATION VIA FILES OF
   STRUCTURES.

2. ADDING CONSTRAINTS - SITE - AND TRANSFORM - SPECIFIC.

3. CONTROL STRUCTURE - RAMIFICATION

   A. ESTABLISH RELATIONSHIPS AMONG PRODUCTS AND REACTANTS
   B. DEAL PROPERLY WITH RANGES OF NUMBERS OF PRODUCTS

4. INTERACTION - DEVELOP MANIPULATION COMMANDS WHICH
                  PARALLEL LABORATORY OPERATIONS, E.G.,
                  SEPARATE INTO FLASKS, TEST CONTENTS OF
                  VARIOUS FLASKS, INCOMPLETE SEPARATIONS,
                  ETC.

5. REPRESENTATION OF REACTIONS

6. PROSPECTIVE DETECTION OF DUPLICATE PRODUCTS BASED ON
   SYMMETRY PROPERTIES OF: A) STARTING MATERIAL; AND
   B) TRANSFORMATION.

*Figure 6.   Major milestones in the development of the
REACT program*

FUNCTIONS

    EDITREACT
    REACT, MREACT
    SEPARATE
    PRUNE

TERMS

    1, E, EX STEPS; STRUCS; PRODUCTS; FLASKS;
    NAMES; LABELS

CONSTRAINTS

    DURING REACTION
       ON STARTING MATERIALS
       SITE-SPECIFIC
       TRANSFORM-SPECIFIC
       ON PRODUCTS

    POST-REACTION
       NUMBERS OF PRODUCTS AT ANY LEVEL
       SUBSTRUCTURAL CONSTRAINTS APPLIED TO
       PRODUCTS (PRUNING) AT ANY LEVEL

Figure 7. Functional commands and descriptive terms used in REACT. EDITREACT allows reaction definition, REACT and MREACT carry out reactions, SEPARATE selects numbers of and flasks for products, PRUNE tests the content of specified flasks for given structural features. Reactions may be carried out as one-step (1), exhaustive (EX), or as equilibrium (E) reactions. STRUCS is the initial list of candidate structures.

product, corresponding substructural constraints are attached to the corresponding flask in the program. As each such assertion is made, the bookkeeping mechanism verifies that, for a set of reaction products from a given starting material, there is at least one way to distribute them among the flasks such that each product satisfies the constraints for its flask.

As an example, consider the problem described above with two products đA and đB. Assume that an oxidative cleavage reaction with appropriate protection had been applied to a set of candidate structures, yielding for each structure a set of products placed in a flask labelled P (illustrated schematically in Figure 8). Separation yielded two identifiable products and perhaps others, placed in the indicated flasks A, B and C (Figure 8). Data acquired on đA and đB were summarized above.

The task of determining legal assignments of products to individual flasks is given to an allocation mechanism. The operation of the allocator will be described in a subsequent publication describing REACT in more detail. However, we can illustrate the problem conceptually with the (purely hypothetical) structures 3-7 in Figure 9. Assuming that 3-7 are candidate structures for an unknown, the statements that the compound đA, in flask A, must have two exchangeable hydrogens and đB in flask B, is a ketone (Figure 8) are constraints on which products of the reaction can be in which flask. Initial separation rejects 3 and its product, đ1 , because only one product is obtained and at least two were required (Fig. 9). Cleavage of 4-7 yields two products đ1, đ2. However, neither product obtained from 4 is a ketone (both are aldehydes), nor does either product possess two exchangeable hydrogens. Therefore neither product can be in either flask and 4 is rejected. Compound 5 yields one product, đ1, which is a ketone and possesses two exchangeable hydrogens. đ1 can be in either flask A or B. But the other product đ2, obeys neither constraint, can be in neither flask and 5 is rejected. Compound 6 yields two products which can be assigned to flasks A and B unambiguously. Compound 7 yields two products each of which possess both a keto group and two exchangeable hydrogens. These constraints are insufficient for unambiguous allocation and either structure may be in either flask.

In the general case, reactions may be carried out in any consistent sequence, whether or not the contents of a flask are uniquely specified. REACT keeps account of the fact that a reaction may be applied to a flask whose contents are known but not assigned to a single structure. Multiple reactions to any level may be carried out. The allocation scheme and ramification mechanism translate statements about products at any level to determine the influence of each statement on the contents of each flask for every structural candidate.

An Example. We have recently reexamined the constraints

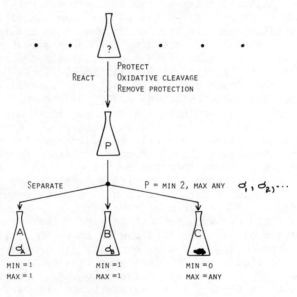

Figure 8. Schematic of the computer representation of a reaction which yields at least two products in the laboratory, both placed in flask P. Flasks A, B, and C result from separation of P into from two to many products.

derived from applications of a dehydration reaction to the structure of palustrol (4). This problem was studied prior to the existence of REACT, so that constraints derived from measurements on the products were translated manually into a substructural constraint on the structural candidates. The data are summarized in Figure 10. Three products were isolated, one of which possessed only a vinyl methyl group (1H NMR), 8, a second which possessed only a vinyl hydrogen, 9, and a third product with neither a vinyl hydrogen nor a vinyl methyl group, 10. The substructure derived from those data was 11, and pruning with this constraint reduced a set of 88 structures to 22 (4). This constraint is incomplete, however, because substructure 11 does not prevent structures in which a methyl group is connected to the methine carbon which does not already bear a methyl group.

A more logical way to do this problem is to carry out the reaction, separation and testing with constraints in REACT and let the allocation and ramification mechanisms decide which structures are legal and which are not. Doing the problem in this way, using substructure 8, 9 and 10 as constraints on the contents of the indicated flasks (Figure 10), reduces the set of 88 structures to 14 rather than 22 because now all implications of the constraints are utilized precisely.

Major tasks mentioned in Figure 6 which remain uncompleted include the problem dealing with incompletely separated mixtures and a general treatment of symmetry properties of structures and transforms. The former problem is difficult because of the combinatorial complexity of the allocation scheme when one cannot assume a single structure per flask (as mentioned previously, however, many possible identities for that single structure are handled properly). The latter problem requires only implementation of procedures for detection of symmetry properties of graphs; procedures which exist in parts of the CONGEN program.

 2) MSPRED. Mass spectrometry is an important tool in organic analytical chemistry. When no other structural information is available mass spectrometry is used as a stand-alone method for structural inference. In combination with other analytical methods it is also extremely useful for post testing possible candidate structures.

Mass spectral data may be used in structural studies in several ways. For example, we can create a fragmentation theory based on examination of sets of known structures and their associated mass spectra. Or, assuming that peaks in the mass spectrum of an unknown originate from unrearranged molecular ions, we can propose possible structures by combining the fragments together under the guidance of a fragmentation theory. When the structure is given we can predict a mass spectrum which will obey the rules of a fragmentation theory. All of these

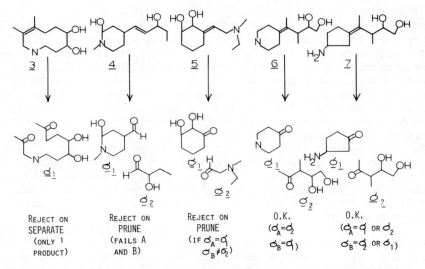

Figure 9.   Candidate structures and results of flask assignment by the allocator under constraints on the contents of each flask (see Figure 8)

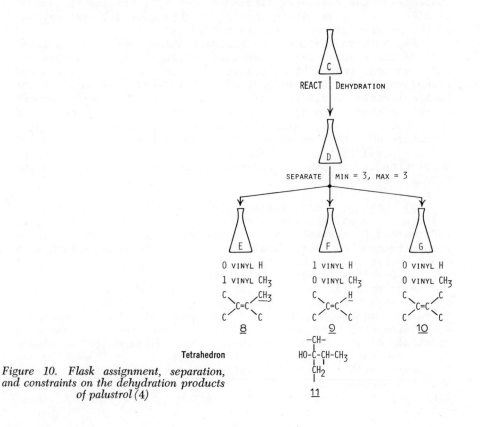

Figure 10.   Flask assignment, separation, and constraints on the dehydration products of palustrol (4)

operations can, in principle be translated into a set of instructions in a computer program. In our DENDRAL programs we have demonstrated mass spectral theory formation when given known structures and the related mass spectra (17,18). A set of fragmentation rules obtained this way were used for studying the fragmentation of estrogens (11). Applying these class specific fragmentation rules, helped us to allocate the position of substituents in unknown estrogenic steroids (19).

When dealing with a new compound (from a class for which mass spectral fragmentation rules are inadequate) for which the only available information is its mass spectrum, we have to use different approaches. If we assume that the fragmentation processes inside the mass spectrometer do not involve significant structural rearrangements, we can attempt to use the mass spectral data for planning. The structures of different ions which are obtained are closely related to the structure of the original compound. We can treat the observed ions as composition (for high resolution mass spectra) or mass (for low resolution) nodes connected together in all possible ways consistent with the molecular formula or weight. The connections are determined by the kind of fragmentation theory we use. We have a preliminary version of a program which performs well for simple test cases but is not yet efficient enough for complicated molecules.

Another use of the mass spectral data is in testing or predicting. Since CONGEN and REACT yield lists of candidate structures which obey spectroscopic and chemical constraints, we can prune further the list of structures using mass spectral information. MSPRED is a program which is a combination of a predictor which uses a theory of mass spectrometry to predict the spectra of candidate structures, and an evaluation function which compares the predictions with the observed spectrum of the unknown, assigning a goodnes-of-fit score to each candidate. The candidates are sorted based upon how well the predicted spectra match the observed. We have examined different theories for prediction; the most useful are summarized below.

a) Half-order Theory. This theory assumes that every bond in the molecule can be broken under certain constraints. In predicting the spectrum, MSPRED explores all possible cleavages of the molecule within these general (user-defined) constraints. Constraints are similar to those discussed previously (17,18), and include limitation of the number of bonds broken and the number of steps in a process, the proximity of pairs of cleaved bonds (i.e. whether or not two adjacent bonds can break in a given process) the multiplicity or aromaticity of each cleaved bond, the allowed hydrogen atoms transfered from or into the charged fragment and the neutral fragments which can be lost. The program calculates the composition and the mass of the fragment which can be obtained in a fragmentation process. The program then combines these results into a predicted mass spectrum with peaks of uniform intensity. The best predictive theories of mass spectrometry are limited to families of closely related structures (i.e., class specific theories). However, given the wide variety of structural types which can be produced by CONGEN and REACT, it is necessary for MSPRED to use this very general model of mass spectral fragmentation.

Evaluation and Ranking. For this general approach we decided to use an evaluation function which takes into account that peaks at high m/e values and high intensity have more diagnostic value then peaks in the low m/e region of the spectrum with low intensity. The simplest form of various evaluation functions we have used is given in Equation 1.

$$SCORE = M_i I_i \qquad\qquad 1$$

where M_i is the mass of a peak present in both the predicted and observed spectra. I_i is the intensity of the (correctly predicted) peak in the observed spectrum.

We expect the half-order theory to be overly complete in the sense that, when applied to the correct structure for an unknown, it will doubtless predict many plausible fragments which are not observed. This simply reflects the fact that the "break everything" approach to mass spectrometry is a considerable oversimplification. Thus the evaluation function does not penalize for predicted but unobserved peaks. What we do expect, though, is that a large number of the observed peaks, particularly the intense ones, will have plausible explanations with respect to the correct structure. Thus a "reward" is given to every observed peak which is correctly predicted (Equation 1).

Example: Our interest in mass spectra of steroids led us to examine a class of mono-keto androstanes as a test case. We obtained the high resolution mass spectra for 10 of the 11 possible mono-ketoandrostanes. These 11 structures were our list of candidate structures. We predicted the high resolution spectra for each of the 11 structures using the half-order theory, and then ranked them against each of the 10 observed spectra. The results are summarized in Table I.

In most of the cases in Table I the correct structure was ranked first and in the remaining it was ranked second. The half-order theory is insufficient to differentiate among mono-keto androstanes when the keto group is located in one of the 4 possible positions in ring A or among structures which are different in the location of the keto substituent in Ring D.

MSPRED is quite new and we have not yet had sufficient experience with it to evaluate its overall usefulness. We are now doing a systematic study of various classes of compounds by ranking the spectrum of a known structure against a CONGEN or REACT generated list of structures which contains the correct structure among several which are closely related. In most of the test cases (including low and high resolution mass spectral data) the correct structure was ranked among the upper ten percent of the structures. We are optimistic that the results of ranking based on the half-order theory can be used as a preliminary filter to divide a set of candidate structures into two portions, one of which has an extremely high probability of containing the correct structure. To this set of top-ranked structures we can apply a more detailed fragmentation theory to make spcific predictions. We are developing computer-assisted methods to make use of a rule-based theory.

Table I. Ranking of Mono-Ketoandrostanes Based on the Half-Order Theory. [a]

STRUCTURE (Keto Position)	RANKING	STRUCTURES WITH THE SAME SCORE	BETTER RANKED STRUCTURES
7	2		6
16	1	17,15	
11	2		12
3	1	1,2,4	
17	2	15,16	2,3,4,1
6	1		
12	1		
15	1	17,16	
1	1	2,3,4	
4	1	1,2,3	

a) Constraints used in predicting spectra included Bonds/Step = 2, #Steps/Process = 1, H Transfers (-2 -1 0 1 2).

2) Rule-based theory. When the candidate structure is known to belong to a previously investigated class of compounds, then we can use additional information to predict a more precise mass spectrum. This information is in the form of specific fragmentation rules. These rules can be used to make predictions of occurrence of ions to supplement or supplant the predictions of the half-order theory. We can also use information about the frequency of correct application of the rule in the set of compounds from which the rules were developed. We call this information the confidence factor associated with the rule. Other important information is the intensity range associated with the peaks which are predicted by a rule. We have the capability of predicting mass spectra using a rule-based theory and have found that by this approach we can

predict a more accurate spectrum and get a better ranking then with the half-order theory. Details of the methods and results of MSPRED using both rule-based and half-order theory will be presented separately.

Summary

We have illustrated a number of approaches to extend the concept of computer-assisted structure elucidation beyond that of simple structure generation. We have illustrated how chemical information together with a computer program can assist chemists in both planning prior to structure generation and, subsequently, testing of candidates. In work described here, the chemist plays an integral part in effective use of the problem-solving tools we provide in the form of interactive programs.

Literature Cited

1. Munk, M. E., Sodano, C. S., McLean, R. L., and Haskell, T. H., J. Amer. Chem. Soc. (1967), 89, 4153.

2. Sasaki, S., Kudo, Y., Ochiai, S., and Abe, H., Mikrochim. Acta, (1971), 726.

3. Carhart, R. E., Smith, D. H., Brown, H., and Djerassi, C., J. Amer. Chem. Soc. (1975), 97, 5755.

4. Cheer, C., Smith, D. H., Djerassi, C., Tursch, B., Braekman, J. C., and Daloze, D., Tetrahedron (1976), 2, 1807.

5. Feigenbaum, E. A., in "Information Processing 68". North Holland Publishing Co., Amsterdam, 1968.

6. Kwok, K.-S, Venkataraghavan, R., and McLafferty, F. W., J. Amer. Chem. Soc. (1973), 95, 4185.

7. Hertz, H. S., Hites, R. A., and Biemann, K., Anal. Chem. (1971), 43, 681.

8. Yamasaki, T., and Sasaki, S., Jpn. Anal. (1975), 213.

9. Jezl, B. A., and Dalrymple, D. L., Anal. Chem. (1975), 47, 203.

10. Schwarzenbach, R., Meili, J., Koenitzer, H., and Clerc, J. T., Org. Mag. Resonance (1976), 8, 11.

11. Smith, D. H., Buchanan, B. G., Engelmore, R. S., Duffield, A. M., Yeo, A., Feigenbaum, E. A., Lederberg, J., and Djerassi, C., J. Amer. Chem. Soc. (1972), 94, 5962.

12. Gray, N. A. B., Anal. Chem. (1975), 47, 2926.

13. Carhart, R. E., and Smith, D. H., Computers and Chemistry (1976), 1, 79.

14. Gatti, G., Cardillo, R., Fuganti, C., and Ghiringhelli, D., Chem. Commun. (1976), 435.

15. Smith, D. H., and Carhart, R. E., Tetrahedron (1976), 32, 2513.

16. Varkony, T. H., Carhart, R. E., and Smith, D. H., in "Computer-Assisted Organic Synthesis," W. T. Wipke, Ed.,

American Chemical Society, Washington, D. C. in press.

17. Smith, D. H., Buchanan, B. G., White, W. C., Feigenbaum, E. A., Lederberg, J., and Djerassi, C., Tetrahedron (1973), 29, 3117.

18. Buchanan, B. G., Smith, D. H., White, W. C., Gritter, R., Feigenbaum, E. A., Lederberg, J., and Djerassi, C., J. Amer. Chem. Soc. (1976), 98, 6168.

19. Smith, D. H., Buchanan, B. G., Engelmore, R. S., Adelcreutz, H., and Djerassi, C., J. Amer. Chem. Soc. (1973), 95, 6078.

Acknowledgment

We wish to thank the National Institutes of Health, (RR 00612 and GM 20832), and the National Aeronautics and Space Administration (NGR 05-020-004) for their support of this research; and for the NIH support of the SUMEX computer facility (RR 00785) on which the CONGEN program is developed, maintained and made available to the nationwide community of users.